ALLE ZEIT WACH
1842

Arie S. Issar

Water Shall Flow from the Rock

Hydrogeology and Climate in the Lands of the Bible

With 51 Figures

Springer-Verlag Berlin Heidelberg New York
London Paris Tokyo Hong Kong

Professor Dr. ARIE S. ISSAR
Water Resources Center
The Jacob Blaustein Institute for
Desert Research and the
Department of Geology and Mineralogy
Ben Gurion University of the Negev
Sede Boqer Campus, 84993
Israel

ISBN 3-540-51621-2 Springer-Verlag Berlin Heidelberg New York
ISBN 0-387-51621-2 Springer-Verlag New York Berlin Heidelberg

Printed in the United States of America

29 00

2132/3145-543210 – Printed on acid-free paper

To Margalit, my wife, who "followed me
in the desert through a land unsown."

Preface

Many times when the author saw the bedouins of southern Sinai excavate their wells in the crystalline rocks, from which this part of the peninsula is built, the story of Moses striking the rock to get water came to mind.

The reader will, indeed, find in this book the description for a rather simple method by which to strike the rock to get water in the wilderness of Sinai. Yet this method was not invented by the author nor by any other modern hydrogeologist, but was a method that the author learned from the bedouins living in the crystalline mountains of southern Sinai.

These bedouins, belonging to the tribe of the Gebelia (the "mountain people"), live around the monastery of Santa Katerina and, according to their tradition, which has been confirmed by historical research, were once Christians who were brought by the Byzantine emperor, Justinian, from the Balkans in the 6th century A.D. to be servants to the priests of the monastery. They know how to discern places where veins of calcite filled the fractures of the granites; such places are a sign of an extinct spring. They also know how to distinguish an acid hard granite rock, and hard porphyry dike from a soft diabase dike. The latter indicated the location at which they should dig for water into the subsurface. In Chapter 9, the reader will find a detailed description of how they used this knowledge to extract water from the rock.

The moral of this story is that when the Sheik of the Gebelia Abu-Heib was asked to explain how water is found in a rock, his answer was "min Allah," which, literally translated, means "from God." Apparently he did not bother to delve into the problem of the flow of water through fractured media, not because it is much more complicated than that of flow in porous media, but because, for him, as for his ancestors and the ancestors of the author, all phenomena, especially those connected with water in the desert, involved the direct intervention of a divine power.

The author received a similar answer when he asked the mullah of an Iranian village in the mountains of Kashan to explain the pulsations of the karst spring irrigating the orchards of the village. A Zoroastrian shrine, still standing above the cave from which the spring emerges, was evidence that the Persian ancestors of the mullah and his congregation also believed that a divine power, and not merely a hydraulic siphon, was involved in causing the spring to pulsate. Whether this power came directly from Ahurmazda, or whether the responsibility was bestowed on a minor divine power, needs further investigation. In the western countries of the Fertile Crescent, the pulsations are attributed to the presence of a gin or genie (a goblin or little devil) inside the spring. Such a spring is the Gihon spring, which was Jerusalem's main source of water and, as will be told later, the water that the Canaanites, and later the Jews of Jerusalem, used for libation.

When speaking about calcite veins in fractured crystalline rocks and about divine powers that affect springs, the author cannot refrain from telling about his experience in Mexico, although Mexico geographically is rather far from the lands of the Bible. In 1981, the author spent his sabbatical in Mexico, collaborating with Dr. J.L. Quijano of the Science Division of the Ministry of Water Resources. The main research being, the isotope hydrology of the groundwater of Central Mexico [1]. While surveying the hills built of volcanics west of Mexico City, a profusion of mineralized calcite veins was observed in the fractures of the basalt rocks. A conclusion was reached, which was later confirmed by the oxygen-18 content of the calcites, that a thermo-mineral spring had once emerged in this place, apparently rather recently, given that the calcites were not weathered.

Dr. Quijano said that he would investigate whether any records of such a spring could be found. A few weeks later, he related that his friend, a historian, had told him that, according to maps and records of the time of the conquest of Mexico by Cortes, such a spring did exist. The Aztec priests used the spring for divination in the following manner. After a priest removed the heart from the bosom of a human sacrifice, the priest would throw the heart into the water of the hot spring. The priest could then divine the future according to the bubbling of the spring's water. Such close interaction between water, faith, and fate can hardly be found in any other place or deed.

Connections, though not so dramatic, between hydrology, water resources, faith, and the fate of people were observed again and again by the author while working throughout the arid countries of the world. These observations have accumulated for about 30 years. The decision to publish them came when the author discovered that many of the traditions and stories, as well as observations, can be explained in a geohydrological conceptual model, making climatic change as a basic key. How and when such a key works will be explained in the following chapters.

The reader should, however, bear in mind that the author is not an archeologist, or a theologist, or an expert in linguistics or in Bible research. He is just a hydrogeologist who believes he is rather well versed in the Bible in its original Hebrew edition. He has also studied the hydrogeological and engineering aspects of ancient water works, especially those in Israel, Sinai, and Persia. He has studied many of the translations of Mesopotamian mythologies and documents in which there were interesting references to water, in some of which he could even follow the linguistic relation between the Akkadian and the Hebrew languages, as both are Semitic. Although he has studied many books on the archeology and religions of the Middle East, still, without false humility, he regards himself as a layman in all these branches of science, except hydrogeology and Quaternary geology. He put to himself the question whether he would rather choose an expert in one of the former sciences, but a layman in geosciences, not to say hydrogeology, to deal with the influence of abundance or lack of water on the fate and faith of his ancestors. As expected, the answer he gave himself was negative, not so much because of his mistrust in the abilities of other scientists to deal with this subject, but because a positive answer would have deprived him of the pleasure of reading the results of the archeological investigations, the study of the history and the religions of the Middle East, and last but not least, of the pleasure he has derived from writing this book. He now invites the reader to share these pleasures with him.

Before the reader begins, some words of explanation are needed in regard to the dates and ages used in this book. When discussing geological and prehistorical events the B.P. (Before Present) scale, had to be used. On the other hand, when dealing with historical events the conventional B.C.-A.D. were, of course, the more appropriate ones. Yet, in proto-historical periods or in the case of historical events to which a geoscientific scale was applied, the author decided to give the two dates. Though the addition or reduction of 2000 to turn one date to the other looks trivial, the author knows from experience that the need to shift from one scale to the other distracts the mind from the trend of reading.

The helpful remarks and suggestions of Mr. Ezra Orion, editor of *Sevivot* (*Environments*), are thankfully acknowledged. Thanks are due to Mrs. Rona Roth and Mrs. Sally Alkon for typing and improving the English and indexing the book and special thanks go to Ms. Jean von dem Bussche for copyediting the book. The professional camera-ready copy was prepared by "Wordbyte", Beer-Sheva. The helping hand and patience, in listening to the lectures by the author on climate change and its influence on history, of the members of the Water Resources Center of the Jacob Blaustein Institute for Desert Research are thankfully acknowledged.

The J. Blaustein Institute for Desert Research
Ben-Gurion University of the Negev
Sede Boqer Campus, Israel

ARIE S. ISSAR

Contents

WHO LAYETH THE BEAMS OF HIS CHAMBERS IN THE WATERS; WHO MAKETH THE CLOUDS HIS CHARIOT; WHO WALKTH UPON THE WINGS OF THE WIND.

WHO MAKETH WINDS HIS MESSENGERS; FLAMING FIRES HIS MINISTERS.

WHO LAID THE EARTH UPON HER FOUNDATIONS, THAT IT SHOULD NOT BE MOVED FOR EVER.

THOU COVERDST IT WITH THE DEEP AS WITH A GARMENT; THE WATERS STOOD ABOVE THE MOUNTAINS.

AT THY REBUKE THEY FLED; AT THE VOICE OF THY THUNDER THEY HASTED AWAY.

THEY WENT UP BY THE MOUNTAINS; THEY WENT DOWN BY THE VALLEYS UNTO THE PLACE WHICH THOU HAST FOUNDED FOR THEM.

THOU HAST SET A BOUND THAT THEY MAY NOT PASS OVER; THAT THEY TURN NOT AGAIN TO COVER THE EARTH.

HE SENDETH THE SPRINGS INTO THE VALLEYS, WHICH RUN AMONG THE HILLS." (Psalms, 104, 3-10)

1 A Hydrogeologist Reads the Bible

> And a river went out of Eden to water the garden; and from thence it was parted, and became into four heads. (Genesis 2:10)
>
> And no plant of the field was yet on the earth and no herb of the field had yet grown: for the Lord God had not caused it to rain upon the earth. (Genesis 2:5)

This book was written by a geologist, whose main fields of experience are hydrogeology and Quaternary geology. It is written firstly for geoscientists interested in knowing not only what happened to the rocks during geological times, but also what happened on their surface during the history of mankind. The author also had in mind the educated layman, whether more interested in the "natural sciences" or whose field of interest is in "humanities". Most of these people are convinced that these two realms should not be mixed. In this book the author will try to merge geohydrology, which belongs to the natural sciences, together with the study of the Bible, history, and archeology, namely fields of knowledge which are classified as humanities.

The book "zooms" into the space-time of the region of the Fertile Crescent during the Holocene. The "guide fossils" are the ancient civilizations which lived in this region in the past, but as with all guide fossils, some of them became extinct, and some went through a process of evolution and survived in their evolved form to the present. In some rare cases in remote places one can still find some "living fossils" which have not changed their ways of life over thousands of years. The book thus contains a synthesis of what the author has learnt about these civilizations and the hydrogeological observations he has collected during the years he worked and travelled in the deserts of the Middle East, from the Sahara in the west, to the Plateau and Gulf of Persia in the east.

The main purpose for his travels was to investigate the water resources of these barren lands. Although as a geologist he was reluctant to take the biblical stories, especially that of the creation of the world, as scientific evidence, he was, nevertheless, impressed by the many correlations he found between his observations and the stories he knew from his readings of the Bible, as well as with the mythologies of the people of Mesopotamia, Canaan, and Egypt. As observations and experiences were collected, the author arrived at the general conclusion that, in most cases, the stories of the Bible reflect various types of experiences which the people of this area went through. These they related to their children in the manner and form in which they understood the world around them. One has only to take into consideration that the further one travels back in the dimension of time, the dimmer and more obliterated the correlations between event and story become. Discussing his findings and impressions with archeologists, the author found that many of them did not agree with his interpretations, especially in all that is concerned to the Pentateuch. They did not regard these books as a reliable source of information, but simply as the miscellaneous remains of a mixture of fables and traditions about past events, that only seldom, and by chance, could be correlated with the findings in their archeological excavations. This disagreement widened later, to an additional field when the author arrived at the conclusion that climatic changes during historical times have played a much more important role in the history of the people of the Fertile Crescent than the archeologists and historians would agree to consider. This conclusion was arrived at in the course of research the author undertook within the framework of an international working group sponsored by UNESCO, investigating the "Impact of Global Climatic Changes on the Hydrological Cycle". This involved paleoclimatic investigations, based on geo-hydrological methods, for information about climates during historical times. Some of the findings showed that there were severe climatic changes during these times. When these were correlated with stories from ancient books, the author found that he could explain additional stories, even those he had until then considered to be just fantasies in the minds of the ancients. One of these was the story of the "Ten Plagues" which were inflicted, according to the Bible, on the people of Egypt just before the Exodus of the Israelites. In this case the author first developed a paleoclimatological conceptual model to explain what happened in this region during cold phases. Then he found that the isotopic composition of a core sample taken from the Sea of Galilee showed that there are indications of a cold climatic phase during the period in which the historians estimate the Exodus to have taken place. He was astonished to discover that the story of the Ten Plagues is more or less his paleoclimatic model described in the words of a man of the second millennium B.C.

As research progressed, the impression was strengthened, that a causal relationship does, in fact, exist between climatic changes and history. The climatic change which had a direct connection with the applied research the author was doing was that which could be seen daily around the campus of The Jacob Blaustein Institute for Desert Research, where the author works. These were the ancient towns and farms deserted about 1500 years ago by their inhabitants. The practical question which was asked was whether the reason for the desertion was due to the conquest by the Arabs or due to a climatic change. As will be explained later, the conclusion of the author was that the "blossoming of the desert" from ca. 300 B.C. to 600 A.D, culminating more or less at the time of Christ, and its desertion later, were the result of a climatic change.

Although the causal relationship between history and climate changes has been suggested by previous scholars, even to the point of explaining all historical events as a result of such changes, this approach has been abandoned by most historians and archeologists except for a few [1,2,3,4]. The opposite school of thought which dominated the world of environmental sciences was that all changes in the environment are a result of the history of man. On the other hand, the conclusions at which the author arrived, through the reinterpretation of pollen and isotope analyses from drillings into the deposits at the bottom of lakes and seas, archeological excavations, etc., caused him to reconsider the exclusive "blame man" approach. He started to look for what he considered to be a more objective evaluation of climatic changes and their correlation with history and even mythology. In recent years, though, the general attitude in the scientific world was starting to change. Putting the blame exclusively upon man was questioned, and scientists have started to lay some of the blame on nature. One of the reasons for this change in attitude is due, in the opinion of the author, to the series of drought and periods of famine which have struck the countries of the Sahel in Africa during the last two decades, a catastrophe, the triggering of which as all observations have shown, had nothing to do with the activities of man, although man helped in aggravating the crisis. The discovery regarding " The Little Ice Age", which lasted from the 14th to the 18th century A.D, and its effects, had also to do with this change of attitude [2,3,4,].

Yet this change in the general approach is taking place slowly, and in general it can be said that the anthropogenic attitude still dominates. This book is part of the effort of the author to try to counterbalance this attitude by suggesting an alternative hypothesis, without claiming that it is the exclusive one.

In this context, the story of how the author came to the conclusion that the anthropogenic approach needs to be challenged is worth telling. This forms the philosophical as well as the empirical basis on which the present book is based.

As previously mentioned, a school of thought, called the deterministic school, was in fashion in the geographical sciences at the end of the last century. The basic theory maintained by the determinists was that all major historical events were a function of climatic changes. The most widely read and cited exponent of this theory was Prof. Huntington from the U.S.A. who, in his books Palestine and its Transformations and Civilization and Climate [5,6] suggested that the simultaneity between climatic changes and major episodes of desertification, migrations, and invasions leads to the conclusion that there exists a cause and effect relationship between climatic changes and historical events.

Sometime during the 1920s, the influence of this school faded out and, during the 1940s, was replaced by a school that preached the opposite theory. It is beyond the scope of this presentation to investigate the reasons for this change; it might have been the destruction caused by man during the First World War, which apparently had no connection with climatic changes. Yet the major impact on the theories of the geographers, pedologists and economists was, in the present author's opinion, the environmental disaster and crisis of the Dust Bowl counties,which took place during the 1930s, when large stretches of land in central U.S.A. were ruined. The blame was put on man who, through his misuse and inappropriate methods of cultivation, brought about the disaster. Today, more and more data points to the fact that this catastrophe was triggered by a series of severe droughts. The high priest of the scientists who had put the blame on man was Prof. Lowdermilk, who, to the original ten commandments, added an eleventh one concerning man's responsibility for the conservation of the soil cover of the earth [7]. This attitude was also the dominant one regarding the history of the Middle East. All around the Mediterranean many ruins of once rich agricultural communities are found, especially in the regions bordering the desert. In the Negev in southern Israel, there are six deserted cities (Fig. 1.1), dating from the Nabatean to the Byzantine periods (ca. 300 B.C. to 600 A.D.), all of which are surrounded by vast areas of terraced valleys, and reveal many wine and oil presses, as well as water cisterns [8,9,10].

While most of the first geographers and travellers who visited these abandoned cities at the end of the last century have laid the blame for the decline of this agricultural civilization on the desiccation of the climate, since the 1930s visiting geographers have laid all the blame on the nomad of the desert, who lay waste to the fertile lands of the Middle East. The main exponents of this approach were the pedologist, Prof. Reifenberg [9], and the botanist, Prof. Evenari who, with his collaborators in their book The Negev [10] claimed that the blooming of the desert of the Negev during Nabatean to Byzantine times was due only to the ingenuity of man in developing methods of water harvesting, while the desertification was due to the invasion of the desert by the Arab nomads, who did not care about agriculture and the maintenance of dams and terraces.

Blame has also been laid on man for disasters in Mesopotamia. There, huge stretches of the country, which were fertile at the time of the ancient civilizations of Sumer and Akkad, are now desolate salt deserts. The archeologists investigating the clay tablets left in the archives of the temples of cities which have since became soil mounds, arrived at the conclusion that from about 2000 years B.C. there were many reports on salinization. They also found that reports on the percentage of barley in relation to wheat in the harvest of the fields increased. As barley was more salt-tolerant than wheat, the conclusion was that man, by his failure to irrigate wisely and drain this valley properly, brought too much water to the area, causing the groundwater table to rise to the surface, the salts to rise by capillary movement, and thus the salinization of the Mesopotamian Valley [11].

During the late 1950s, the author discovered that all the sand dunes covering the coastal plain of Israel (Fig. 1.1) were deposited after the termination of the rule of the Byzantine Empire, namely during the rule of the Arabs over the Levant (from ca. 670 A.D.). Being influenced by the anthropogenic doctrine, the author suggested in his Ph.D. thesis [12] that the conquest of this land by the Arabs was the cause of the sand invasion. His explanation was that the conquest by the nomads led to the destruction of the agricultural plantations as well as all the remaining natural vegetation, which prevented the advance of the sand dunes.

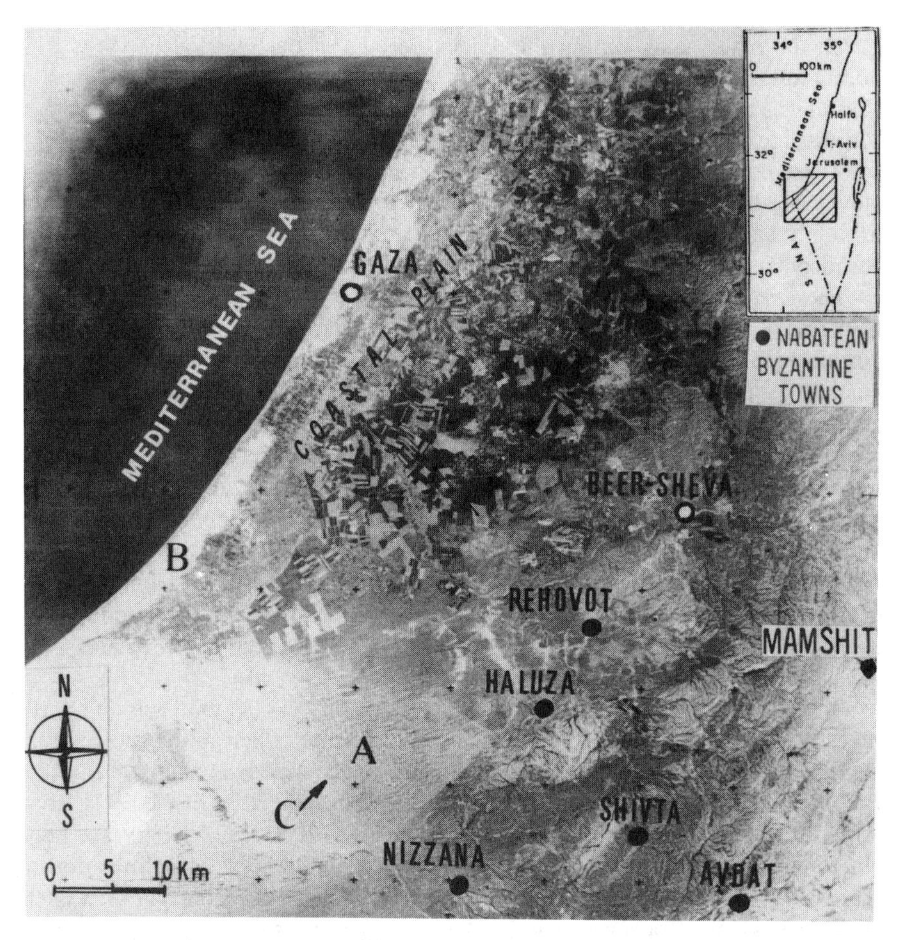

Fig. 1.1. Satellite image of southwestern Israel. A. Dunes of Lower Holocene age. B. Dunes of Post-Byzantine age. C. Border between Sinai (Egypt) and Israel. Over–grazing effects can be seen on the Egyptian side

The author began to question his own conclusion when his paleohydrological and paleoclimatological investigations in the Negev and Sinai taught him that the sand invasion was due to climatic changes. He found that, during the periods of glacial retreat, the global arid belt of the Sahara moved northward, causing the Levant to become dryer; this movement, apparently, caused the Ethiopian mountains to be subject to monsoon rains. The floods sweeping down from the mountains and through the plains of Nubia brought the sands into the Nile, which then took it into the Mediterranean Sea, where the currents and waves carried it to the coastal plain of Israel.

On the other hand, during the glacial periods of the Pleistocene, heavy dust storms followed by rain storms caused the deposition of loess, while the deposition of sand was at a low level (Fig. 1.2). The question then occurred to him whether it is not possible that the same might have happened at the end of the Byzantine period before the conquest by the Arabs, namely, that due to natural reasons the desert belt moved northward causing desertification of the Levant. During this same period Ethiopia came under the dominance of monsoons. This could explain the abandonment of the desert cities, as well as the invasion of sand.

The first occasion to expound this theory was at the annual meeting of the Quaternary geologists of Israel. At the same meeting a study of the pollen assemblage found in a core sample taken from the Sea of Galilee was presented by a young palynologist, who discovered that during the period of the Roman-Byzantine empire the ratio of olive pollen increased while that of oak and pistachio of the natural forest decreased. At about 500 A.D. the ratio changed drastically, the ratio of olive decreasing and that of the natural forest increasing [13,14]. The explanation given by the palynologist and his colleagues, the archeologists, to this environmental change was anthropogenic, namely, that the olive plantations flourished while the governments supported the farmers, at which time the people cut the natural vegetation to be replaced by olive trees. The plantations were abandoned when they were overtaxed by the Roman government. This enabled the reestablishment of the natural vegetation.

The present author was not ready to accept this explanation without further investigation. And indeed, a closer study of the same pollen ratios showed that a similar change in the ratio of olive and oak pollen occurred at about 4500 B.P (2500 B.C.) (Fig. 1.3). This was also a time when the Canaanite cities of the Early Bronze Age of the Negev flourished. At about 4000 B.P. (2000 B.C.) a decrease begins which is also the time when in Mesopotamia indications of salinization and increasing ratios of barley relative to wheat became evident. There was thus additional evidence for suspecting that the invasion of the sand dunes and the increase and decrease in the pollen of the olives and the rise in the barley ratio was indeed the result of a climatic change rather than of man's abuse of nature.

Fortunately, additional evidence, which was not at the disposal of the earlier scientists, became available. This was the results of the environmental isotopes such as oxygen-18 (^{18}O) and carbon-13 (^{13}C). These are not influenced by man's intervention, and are therefore objective witnesses to what happened in the past. It happens that the same cores from the bottom of the Sea of Galilee also furnish oxygen-18 evidence, which shows that, during the periods of high olive ratios and wheat production, the ^{18}O and ^{13}C ratios were low, which is typical of colder climates (Fig. 1.3) (The depletion of ^{13}C is due to a higher ratio of biological activity in the soil. Regarding ^{18}O see Appendix II). The results of these studies were presented at international scientific conferences and published in their proceedings [15,16]. The reaction was not long in coming; colleagues came forward and volunteered information which supported the hypothesis of climate change. Examples are the information on the post Roman sea transgression over the coasts of England, which was sent by the archeologist Mr. Brian Yule [17], that of the western U.S. supplied by Dr. Owen Davis from The University of Arizona [18], that of the Swiss Alps, and that of the speleothemes from Israel by Prof. Mebus Geyh of Hannover [19]. Of special interest was the incident at NATO's Advanced Research Workshop on Paleoclimatology and Paleometerology in Oracle, Arizona, in 1987, when the author was asked whether he had any observations about an increased amount of dust deposition during the Roman period, as this should be the result of a colder climate, the same that occurred in this region during the glacial periods of the Pleistocene. On the author's answer that he had no direct evidence to support such a conclusion, Dr. Bücher from France presented his observations on the rate of dust storms in France and one of his diagrams summarized all the historical evidence, and, in fact, an increase in the rate in dust storms could be seen for the period in question [20]. Part of these data were put together in one diagram. (Fig. 1.3), and the reader can see that a rather good correlation exists between historical events and major climatic changes. The question which still has to be answered is whether this is a coincidence or a causal chain of events.

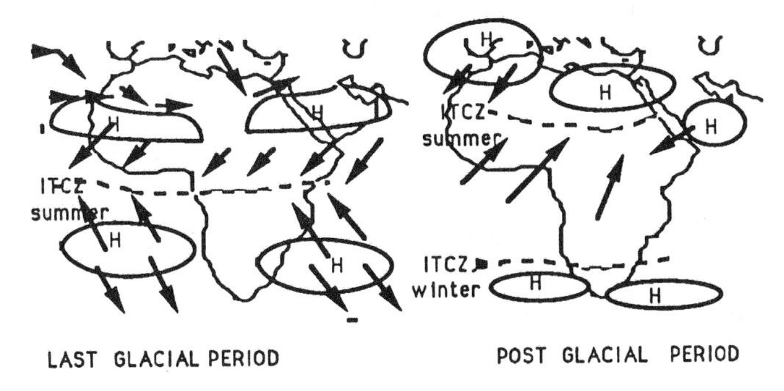

Fig. 1.2. Paleoclimatic maps of the Mediterranean and North Africa during the Last Glacial period and during the beginning of the Holocene

Unfortunately, the ancients did not keep records on temperatures and precipitation, and thus they left enough room for debate and argument regarding the level of impact which climatic changes had on historical events. Yet it is difficult to argue against the possibility that climatic changes, once they had been found to have happened, simultaneously with the migration of people, desertion or flourishing of cities etc., had an impact on history, culture, and society. At the same time one must also take into consideration the possibility that under certain circumstances man's activities may influence the environment. This may happen when overgrazing takes place, or when man cuts down vegetation. This will cause the land surface to become brighter, which causes higher back radiation (higher albedo). This causes the surface to become cooler than the air above it, which prevents upward movement of air masses, namely convectional air currents, which may cause rain [21]. In addition to this, other anthropogenetic changes occur which involve erosion, salinization, etc. The effect in many cases may be self-accelerating due to the fact that in times of want man may over-exploit the resources of the land. As a result of all these observations and considerations, the author does not suggests adopting a strictly deterministic or anthropogenic approach, but to regard nature as a dynamic system, composed of two subsystems, man and environment. Influences move in all directions; changes in the natural environment influence culture and history, which then reciprocate and influence the environment.

In the special case with which this book is concerned, namely the story of the lands of the Bible, the reader must take into consideration that this region is characterized by its aridity and thus the dependence of man on rain and spring and river water, and that periods occurred when these resources became depleted or too abundant. In this work, events in which water played a role will be surveyed, and the special hydrological and geological conditions which might have caused them will be described. In addition to this, the environmental conditions during the times when these events took place will be reconstructed. In the opinion of the author, all this will give the reader a better and deeper understanding of ancient texts and, as was the experience of the author, may add a new flavor to the reading of these books.

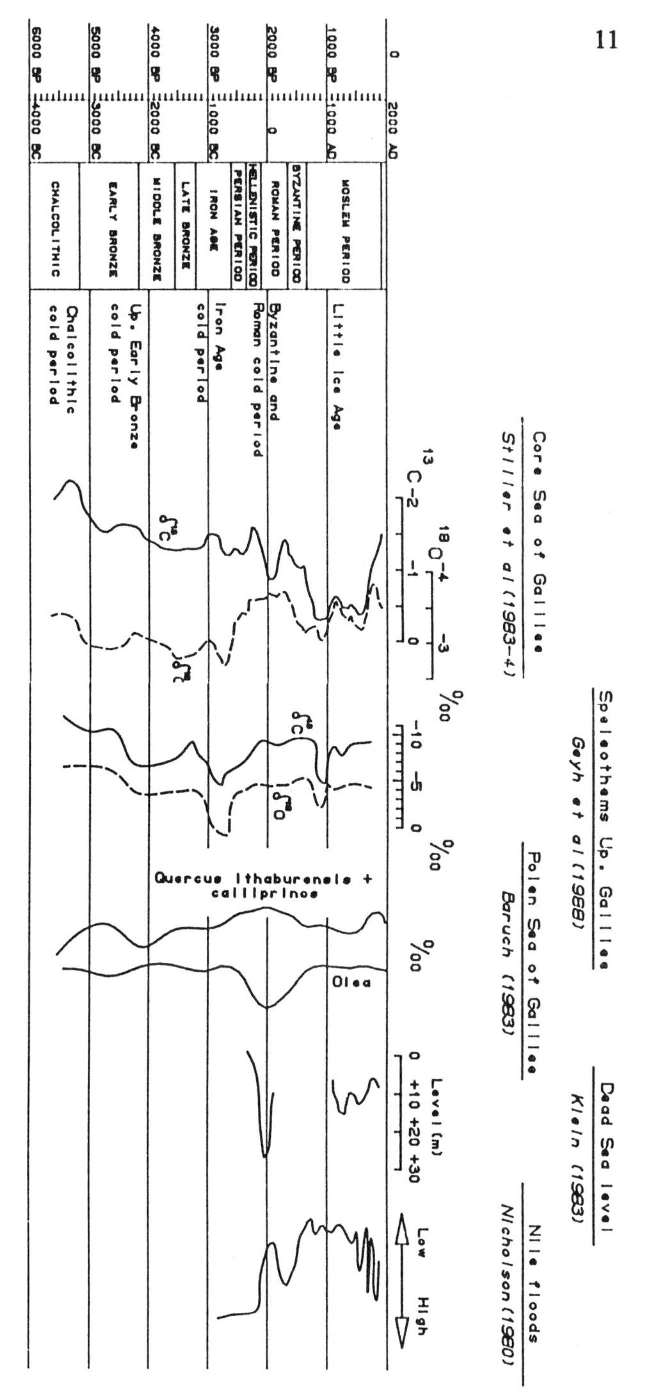

Fig. 1.3. Paleoclimatic correlation diagram

2 Rocks, Water, and Soils

> And God said, Let the waters under the heaven be gathered together unto one place and let the dry land appear; and it was so. And God called the dry land earth and the gathering together of the waters He called the seas; and God saw that it was good. (Genesis 1:9-10)

The name "earth" in English or "Erde" in German, by which the dry land was called and by which we name the globe we live upon, is "eretz" in Hebrew, which is the language of the Bible. In Akkadian it would have been "Ertzitu", in Aramean "area". The word "fruit" in the same chapter of Genesis, which is "Frucht" in German, is "pri" in Hebrew, and "pr" in Phoenician. The reader may find a further similarity between the "wine" he drinks and the "yain" which Noah enjoyed, in Chapter 9 of Genesis, while relaxing after he spent 40 whole days and nights in one ark, only three hundred cubits long and fifty cubits in breadth, with representatives of all the animals of the world, in addition to all his family, drifting on the waters of the Deluge.

Other similarities may be found between "wheat" and "hitta", and "plough" and "fellah", which means a peasant in Arabic, and "paloah" which in Hebrew means to cut through. Another similarity is between a cotton coat and "cotonet passim" which Jacob made for his son Joseph (Genesis 37:3).

It is not the purpose of the author to indulge in a philological exposition. These examples were given in order to show that information, especially that concerned with agriculture (not to speak of writing) in ancient times was transferred from east to west and north.

The main agents for this transfer of know-how were most probably the Phoenicians, who spread their colonies throughout the Mediterranean area and traded even beyond it. Some exchange might have been made through the Greeks and the Romans, who ruled the east for many centuries. The geographical term "Middle East" or "Near Orient" was coined by the people living west of this region. To the people who were living there before civilization spread west and north, this region was the center of the earth. The Sumerian believed that the world came into existence in Sumer, where the gods built and chose their cities to dwell in and, hopefully, to protect them. The Akkadians (Babylonians and Assyrians) believed that Babylon was the gate of heaven. The Hebrews believed, and still believe, that Jerusalem is the place where the Almighty chose to build his Temple. Christianity adopted this belief and many ancient maps show Jerusalem as the center of the world. The ancient Egyptians had no doubt, even for one moment, during their long history that the emergence of the Primeval Mound from the Primeval Waters was in the land of the Nile, most probably at the site of the temple of Atum-Ra at Heliopolis.

Thus, for the sake of historical objectivity, it is suggested that the region in which the events narrated in the Bible took place should be referred to as the "Fertile Crescent" in a very broad sense, namely to include Iran, Mesopotamia, and Egypt.

The beliefs concerning the center of the world which were written on clay tablets, engraved on stone, or written on papyrus or parchment, were later covered by the ruins of magnificent cities, buried in deep tombs, while the centers of civilization moved gradually to the west and north, first to Athens and Rome and later to western and northern Europe and to America. The scientists who started to unearth the ancient sites came mainly from Europe and the United States, although the roots of their sciences began from the civilizations they were looking for. The return to the east was due to the general scientific endeavor to better understand the history of humanity. Another urge was the wish for a deeper knowledge of the roots of Christian civilization, especially all that is connected with the stories of the Bible. The archeologists who excavated the "tels" (mounds) and cemeteries in the Orient constantly looked for a correlation with stories narrated in the Bible and those of classical Greek writings. Little effort, however, was invested by the archeologists in searching for a correlation between the archeological findings and the hydrology and groundwater resources of the region, not to speak of considering a different climate in the past to explain their findings.

As stated, this type of information was found by the author to be of importance for constructing a comprehensive historical picture, but before embarking on the voyage back in the dimension of time, an introduction into the physical environment of the region is suggested.

The reader who is not a geoscientist may skip this introduction, as a briefer description of the special character of each subregion will be given at the beginning of each chapter.

The region in which most of the stories told in the Bible took place is characterized as being a transition zone in all possible aspects: geological, climatic, ecological, as well as socio-cultural. It is argued that the primary factor is the geological one, which is decided by the fact that the region is a border zone between plates, namely, the Arabo-Nubian and that of Euro-Asia with the Tethys in between (Fig. 2.1.). The Arabo-Nubian shield, which is exposed in the most southern part of the region, extends further southward to Africa and Arabia, and is the northern part of an ancient continent, Gondwana, which has been broken and rebroken into plates which have moved from one place to the other since Paleozoic times until they settled into the present configuration of the continents [1].

The Tethy sea extended to the north of this continent, and also changed its shape according to the movement of the continents as well as due to other tectonic movements which affected the face of the earth beneath this sea and on its margins. Such movements brought changes in the extension of the sea which were expressed in transgressions and regressions. These movements decided the nature of the sedimentary column throughout geological history.

Another factor which influenced the tectonic pattern of the region and caused its uniqueness is the huge fracture system in the surface of the earth, called the Syrian-African rift system. This rift crosses the entire region, running from north to south. It was and still is a major factor in the evolution of its geological as well as morphological character (Fig. 2.1).

Before plunging into the Tethys sea and Syrian-African rift, one has to go back a few hundred million years in time to understand the origin of the basement rocks, crystalline as well as metamorphic, belonging to the Pre-Cambrian era, which are found mainly in the southern part (Figs. 2.2, 2.3).

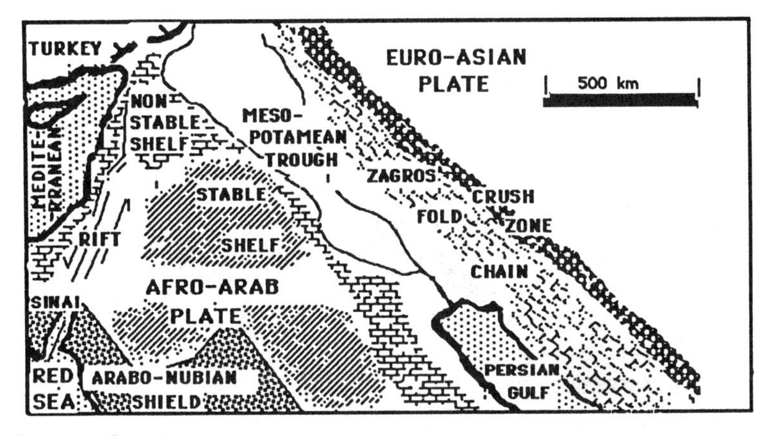

Fig. 2.1. General structural map

Originally, the most ancient rocks were sandy and silty sedimentary layers which, due to high pressure and temperatures, were changed into mica schists, gneisses, and other types of metamorphic rocks. These were penetrated by hot magmatic intrusions, which ranged in chemical composition from basic magmas like black gabbro nurites to acid rocks of the granite types discerned vividly by the red color of the rocks on the eastern side of the rift in the mountains of Edom ("adom" in Hebrew means red).

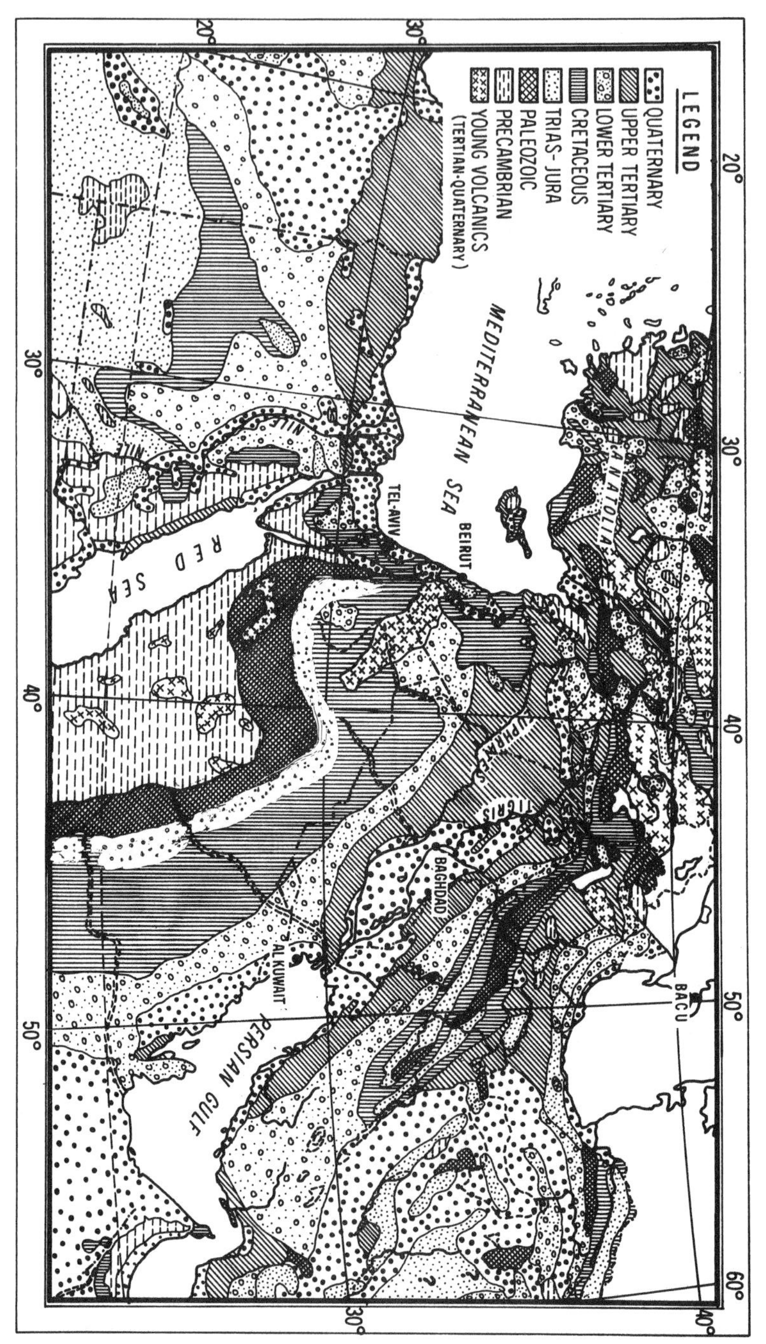

Fig. 2.2. General geological map

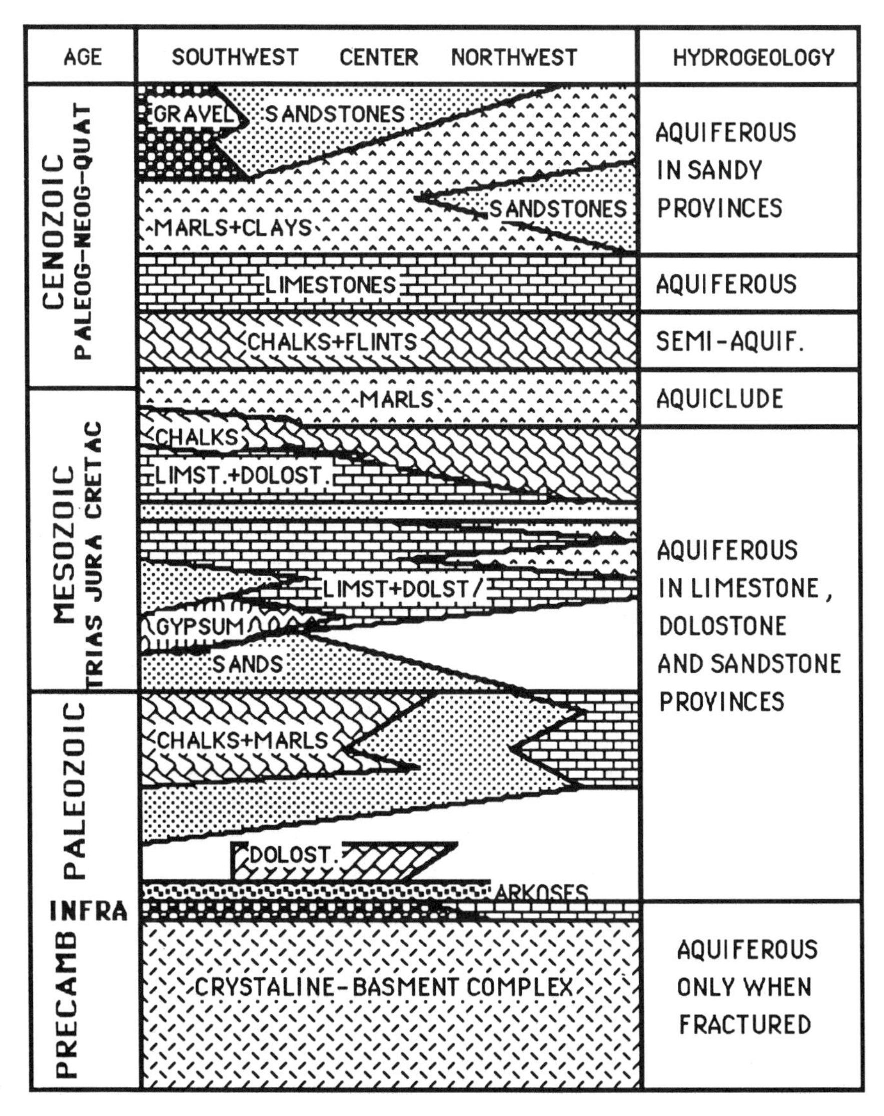

Fig. 2.3. Geological correlation chart

The intrusion of these plutonic bodies was a prelude to an extensive tectonic upheaval which created a very high mountain range all over the area,which at that time might have been of the order of magnitude of the present Alpine mountain ranges.

Erosion forces which worked on the rocks forming these mountains turned them into debris which was washed down into the valleys. There it was penetrated and covered by lava flows which emerged from fractures and volcanos, giving vent to the high pressures and temperatures accumulating in the interior of the earth.

In conclusion, the complex of Pre-Cambrian rocks at the southern part of the region represents one of the earliest geological phases of sedimentaion folding, faulting, and magmatic intrusion, in the following order:

1st stage: A sedimentary basin of an oceanic island-arc nature which went through phases of subduction and collision. In this ocean flysch, sands, and clays were deposited a few thousands of meters thick. While being deposited, these layers were penetrated by basic intrusions.

2nd stage: A strong folding event which caused extensive metamorphism followed by magmatic intrusions becoming more and more acid toward the granitic type of rocks. An erosion of the folded mountains formed an unconformity between the metamorphic rocks and those that succeeded them.

3rd stage: A regional upheaval along regional fault lines where volcanic rocks erupted.

4th stage: A regional phase of erosion turning the high mountains into a regional plateau. The products of the erosion of the high mountains, together with the lavas and other volcanic rocks, filled the valleys and left behind rolling plains to be covered by the thick sequence of sandstones.

During all the phases of Pre-Cambrian faulting and folding, the area was penetrated by dikes of various types, from coarse-grained ones forming quartz pegmatites to fine-grained porphyrys, quartz porphyrys, and diabases. These dikes criss-cross all the columns of Pre-Cambrian rocks and thus give the area its special morphological character. The more resistant dikes, mainly the quartz porphyrys and porphyry, were less affected by the erosion than the quartzitic or metamorphic rocks into which they penetrated. They protrude as long natural barriers on the surface of the land. On the other hand, the more basic dikes, such as the diabase and gabbros, were less resistant than the granitic rocks into which they penetrated and, due to this, they form natural ditches. (The protruding dikes are called positive, while the weathered ones are called negative dikes). They are of importance in regulating the subsurface groundwater flow from the fractures of the granites. This subject will be discussed in more detail in Chapter 9, dealing with the water resources of the Sinai Peninsula.

At the end of the Pre-Cambrian, a long hiatus in the depositional record took place. During this hiatus, the surface of the land was flattened to form a peneplain extending over all the ancient continent. This peneplanization process marked the termination of the era of tectonic activity of the Arabo-Nubian shield. From the end of the Pre-Cambrian onward, this shield was not subjected to any further folding. It did, however, undergo slow, up and down movements (epeirogenic) as well as breaking-up processes followed by volcanic eruptions and earthquates along the fracture zones. On the whole, the crust of the earth became stable and the main geological processes which started on the surface of the shield became those of erosion and deposition. The first caused the rocks to disintegrate into components, while transport and depositional processes, water and wind working as agents moved the products of disintegration to the lowest places and piled them up to form layers of various types of rocks. These processes first started by the collapse of rocks under their own weight when the bonding of their particles was weakened by physical or chemical processes such as freezing and unfreezing or by dissolution. Afterwards the rocks were broken into even smaller particles while they were transported further. Another agent, the wind, worked either on the face of the exposed rocks or on the disintegrated particles, picking up the lighter ones and carrying them away during dust or sand storms, or rolling them or having them "jump" from one place to another depending on the amount of wind force.

These processes of disintegration, erosion, transport, and deposition in basins lasted for many millions of years until the once mountainous Pre-Cambrian landscape became flattened. Highlands became plains and valleys filled up (Fig. 2.4).

After the peneplanization was complete, a process of deposition started to take place. This brought about the building up of thick layers of conglomerates and sandstones on the flat surface of the crystalline rocks throughout the northern margins of the Arabo-African shield, extending over Arabia, Sudan, Jordan, Israel, Egypt, and the Sahara to the coast of the Mediterranean Sea. The sequence of sandstones is built, in its lower part of an arkose, of coarse sand and pebbles of feldspar and quartz. Toward the upper part, the sand becomes more composed of quartz, finer, and well-sorted. The sands were impregnated by various minerals which were dissolved in the water depositing the rocks, or even after their deposition. These minerals contained iron, manganese, and copper and rendered various colors to the sands, red, black, and green. These colors, together with the depositional patterns of bedding, cross-bedding and the interfingering of clays and silts, give the desert landscapes in which these rocks are exposed a special colorful character. The copper minerals have been mined since the Bronze Age by the people dwelling in these regions.

The sandstones were deposited on the continent, and only seldomly did the sea encroach upon it to deposit layers of marine origin. This fact makes the dating of these rocks very difficult, as in a continental environment fossils are rather rare, due to poor preservation conditions. Rocks on which isotopic age determination can be done are also rare. Thus the ages of the sandstones are poor and vaguely established, and only where there is an interfingering of marine layers or some aquatic deposits containing fossils, or volcanic rocks on which isotopic methods can be applied, is dating feasible. In general the sandstones are porous and are important as groundwater aquifers throughout the region.

One of these transgressions occurred during the Cambrian, when the sea deposited silt, dolomite, and limestone layers, the nature of which portrays an environment of a shallow sea and lagoons which were sometimes rather brackish.

The shore of the continent was situated at the present southern extension of the Red Sea, namely, the Indian Ocean, but the sea extended to the north. All the region, sea as well as continent, were in the extreme southern part of the globe. Glacial deposits of early Paleozoic which are found in central Arabia are evidence of the proximity of this region to the southern polar glaciation region [2]. During the

Ordivician and Carboniferous the sea transgressed again. The sea came from the northwest and deposited limestones, dolomites, and shales. During the late Permian another transgression deposited carbonate layers in eastern Arabia and Iran. At the end of the Paleozoic the breaking-off of the Gondwana continent from Pangaea, the Pre-Cambrian primordial continent, began.

During the Mesozoic the continent drifted northward; it passed from the low to the middle latitudes, its margins were again and again covered by the ocean, in which mainly carbonates were deposited. The first trangression over the margins of the new continent was during the Triassic.

Fig. 2.4. The Pre-Cambrian peneplain in southern Sinai as seen on a NASA satellite image

In the cirque of the Ramon, (a geological park in the southern part of Israel) the sequence of layers of this transgression reveal all the stages of a marine ingression. In the lowermost part one finds silty, clayey deposits typical of a seashore environment. In these layers bones of marine reptiles can be found. On top of these, one finds limestones and clays rich in sea shells. These are overlain by layers of gypsum, denoting the regression of the sea and the development of closed lagoons .

The regression of the mid Triassic sea was most probably due to upheaval and tectonic movements which can be detected according to unconformities at the end of this period.

During the Triassic period, the breakdown and movements of the plates brought the center of the region into the vicinity of the equator, and when continental conditions were reestablished, the tropical climate caused the formation of red lateritic soils like those forming today in the tropics.

At the beginning of the Jurassic period, the sea encroached again, depositing limestones and marls, the deposition continuing on the shelf of the Tethys sea during most of the this period. These limestones deposits formed the Lebanon and Anti-Lebanon mountains, their high permeability, due to karstification processes, make them good aquifers. The great springs flowing from these rocks help to irrigate the Damascus basin in Syria and also form the sources of the Jordan river. At the end of the Jurassic period a regional tectonic upheaval took place. This upheaval was accompanied by magmatic intrusions which penetrated the Jurassic rocks as dikes, sills, stocks and lacolites, followed by flows of basalt. The regional upheaval caused the sea to regress far to the north and the layers of the Lower Cretaceous period is of a typical continental facies of sandstones known throughout the region and from northern Africa .

The post-Jurassic upheaval and Lower Cretaceous regression in many places caused the erosion of the Jurassic rocks and brought an unconformity between the rocks of the Lower Cretacaous and the Triassic.

As will be discussed later, the sandstone rocks of the Lower Cretaceous are also very important as water-bearing strata in the deserts of Arabia, Jordan, Israel, and Egypt. This is due to their porosity, as well as to their continuous distribution over extensive parts of the region.

The Middle Cretaceous marks the end of the dominance of continental conditions, the sea encroached and covered a big part of the continent during most of the time up to the Neogene. This sea was already the Tethys sea, namely, the precursor of the Mediterranean, as at the end of the Jurassic the main continents were already divided in the general configurations known to us today. The sea trangression came from the north, thus as we go northward from the ancient continent, the marine sediments become thicker and of a deeper sea facies.

The rocks deposited during the Middle Cretaceous over most of the region were mainly limestones, rich in marine fauna such as ammonites, and rudists. These rocks are hard and resistant to erosion and most of the highlands of the region are built of them. After the sea regressed these limestones went through a dissolution process ("karstification") and became highly permeable. This gave rise to the formation of a thick water-bearing sequence (aquifer), which is able to store very large quantities of water, flowing out from large springs, which lead into the rivers.

At the end of the Mid-Cretaceous, the entire region went through a folding phase which brought about the formation of anticlines and synclines, the top of which protruded above the sea. This was most probably connected with the collision of the Afro-Arabian and the Euro-Asian plates with the smaller plates of Turkey and Iran between them. This movement also caused the closure of the Tethys sea and the formation of the Taurus and Zagros mountain chains (Fig. 2.1).

At the end of the Cretaceous, the sea-land configuration in the region was that of narrow bays located in the synclines penetrating between the peninsulas and islands formed by the anticlines.

At the beginning of the Cenozoic the sea started again to encroach on the land, depositing marls and chalks rich in calcareous globigerina and siliceous radiolaria ooze. The abundance of silica in the sea water during the Lower Eocene also caused the formation of flint layers which were interbedded in the chalks. At the Mid-Tertiary (mainly during the Middle Eocene) the sea covered most of the regions. In this sea, limestones and chalks were deposited. At the end of the Middle Eocene, the sea regressed and a continental elevated plateau formed in the south, from which still higher mountain ridges emerged. This period was that of the formation of the regional anticlinoriums extending from northern Sinai to northern Syria, including the mountain chains of Israel and Lebanon. The general inclination was toward the Mediterranean Sea and streams which joined rivers started to erode the highlands and transfer the eroded material from the high mountains to the low basins and sea. Some synclinal regions were filled up by the continental deposits. Thus the Upper Tertiary, mainly the Neogene, is characterized, on one hand, by featuring the landscape through erosion, forming the main land forms of the region. On the other hand, the deposition of thick continental sediments as well as that of shallow sea sediments in big basins took place. These deposits are composed of fresh and brackish water carbonates, conglomerates, namely, pebbles and gravel consolidated by calcium carbonate, sands and silts.

During the Neogene the Syrian-African rift system started to form along an ancient suture line (possibly Pre-Cambrian), separating Africa from Arabia. The fracture system forming the rift valleys of the Fertile Crescent is a strike slip slope (Fig. 2.1), associated apparently with the opening of the Red Sea and the breaking away of the Arabian sub-plate from the African one [3]. The bifurcation of the rifting system also separated, during the Lower Pleistocene-Upper Pliocene periods, and the peninsula of Sinai from the neighboring plates.(Fig. 2.4). The tectonic movement accompanied by seismic activity along these horizontal faults continues to the present. This movement also had an effect on the neighboring geological structures; it caused them to break into blocks, tilt and rotate. This apparently caused the structural configuration of the mountain chains which formed the highland backbone of the Fertile Crescent (Fig. 2.1). The rift valleys were filled by thick continental deposits. This was accompanied by the outflow of lava along fracture lines, forming thick deposits of basalts .

During some short intervals the sea again penetrated, but this time it covered limited areas. It entered into the syncline zones, depositing layers of limestones. In the Upper Neogene (Late Miocene) an extreme regression of the sea caused the Mediterranean to dry up and deep canyons were formed, draining into the nearly dried-up Mediterranean basin [4]. The retreating Mediterranean Sea left behind salty lagoons in the bays and rift valleys in which salt layers were deposited, as well as lake deposits containing remnants of freshwater fauna.

The severe tectonic movements which were responsible for the formation of the valleys of the Syrian African Rift Valley during the Uppermost Neogene and Lower Quaternary caused these layers to become severely folded. Due to pressure, the salt became fluid and folded upward as big salt plugs, one of which is Mount Sodom.

As mentioned, during this period the entire region went though a renewed tectonic uplift apparently connected with the shear movements along the Syrian African rift system. This caused an upheaval in the backbone of this region, thereby forming the anticlinorial ridges stretching from Israel to northern Syria. At the same time, the rift valleys developed further, being filled by the eroded material from the rivers, which, until then, had crossed them flowing from east to west. In the synclines, conglomerates and silts were deposited by rivers carrying products of erosion from the surrounding mountains.

At the beginning of the Quaternary the physical background was set up for the arrival of man on the land. During this period the landscape, as we know it today, was finally formed. During the Pleistocene period, climatic changes from glacial to interglacial ranged over Europe, while in this region pluvial and interpluvials occurred. During the Lower Pleistocene, the climate was humid, springs flowed from the limestone aquifers into rivers which flowed from the highlands into the synclinal basins where gravel and muds were deposited. In regions where springs emerged directly from the limestone aquifers into the basins, the calcium carbonate dissolved in the water precipitated, hardening the gravel into a conglomerate. This deposition of gravel and silts is evidence of torrential flows in the rivers, most probably due to lengthy rainstorms. In the basins black and brown alluvial soils were formed, while on the limestone outcrops typical red soils of the "terra rosa" type were formed. The formation of these soils occurs under Mediterranean climate conditions and are often referred to as Mediterranean-type soils [5,6].

The limited penetrations of the sea during the Pleistocene are correlated with interglacial periods. The rise in sea level was caused by the vast quantities of water released by the melting of the glaciers. There exists ample geological, as well as prehistorical, evidence for a few pronounced changes in the climate of the region during the Pleistocene. As already mentioned, some of these changes can be correlated with the glacial and interglacial periods which caused global changes. The effects on desert regions were humidification during the glacial period, and aridization during the interglacial periods. The most documented are the humid period, in correlation to the Last Glacial period, and the aridization which followed at the termination of the Pleistocene and the beginning of the Holocene.

During this period of the Uppermost Pleistocene and the transition to the Holocene, there were undoubtedly also changes in the location of the Sahara belt, as evidence from subtropical Africa shows that during the Last Glacial period levels of lakes and terraces of regional rivers were low and rose at the end of the Pleistocene. Yet the correlation between humid glacial periods for the earlier Pleistocene as well as minor climatic changes during the Holocene is not yet well established.

The primary factor deciding the availability of water resources is the climate. The Fertile Crescent, as all Mediterranean countries, is influenced by the Westerlies Zone, bringing in rainstorms from the northern Atlantic during winter. During summer it is influenced by the Sub-Tropical or Inter-Tropical Convergence Zone overlying the deserts of Arabia and the Sahara. In this convergence zone, air masses, which ascend in the tropics, descend, and while doing so, become heated and thus relatively dry (Fig. 1.2).

The movement of these belts northward during the summer season and southward during the winter is due to the change in the position of the sun as a function of the inclination of the earth and its path around the sun.

During the winter, the dry climatic belt, characterized by high barometric pressure, moves south, the Mediterranean and surrounding countries are then influenced by a belt of low barometric pressures bringing moist air and rains from the Atlantic and the North Sea.

The movement rate of the belts southwestward and thus the number of rainstorms reaching the region is different from year to year. In years when the belt of high pressure remains over the area, the rainstorms are less abundant and the year is dry.

The impact of these changes, which influence the whole of the Fertile Crescent, is crucial in the zones bordering the desert. Moreover, due to the configuration of the coastline of the southeastern edge of the Mediterranean Sea, the deserts of northern Egypt, Sinai, Negev, and southern Jordan lie outside the main path of the rainstorms that come from the west.

These two factors, namely, the southern desert belt and the trajectories of the marine storms, affect the mean annual quantity of rain as well as its variability from year to year. As can be seen from the multi-annual precipitation map (Fig. 2.5), precipitation becomes less as one goes south and east. Yet, one also has to take into consideration the topographical factor; thus, while the rift valleys are in the shadow of the rain coming from the sea, and relatively arid, the mountains receive more rain and snow in winter.

The scarcity of rains, on the one hand, and the high variance, namely the difference from year to year on the other hand, become more and more severe as one goes further into the desert. It can be said that the rains in the desert are characterized by scarcity and randomness.

The rains come as winter rains between the months of November and March, and, in this respect, the region is fortunate in relation to other regions, where the rains are summer rains. The temperatures during the winter are relatively low, thus the relative effect of the winter rains is rather high, as evaporation when the rains fall is relatively low. Many of the rainstorms affected by a barometric low in the northern and central part of the Fertile Crescent come to the desert areas as smaller eddies on the margins of the bigger cone of barometric pressure. They form small convective cells, a few to tens of kilometers in diameter. When this occurs, rain falls on a

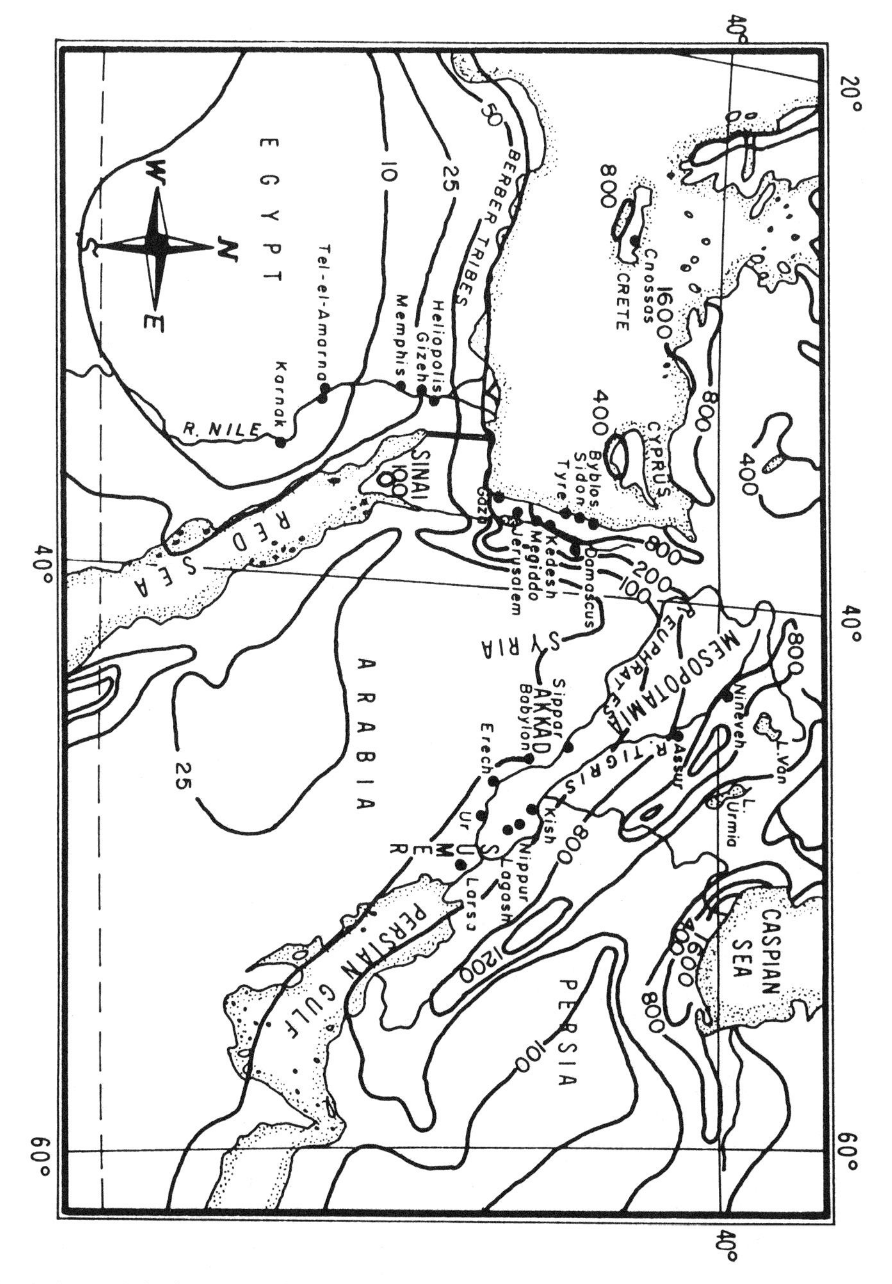

Fig. 2.5. Precipitation map of the Fertile Crescent (based on [7])

limited area around the center of the cell, while other areas may remain dry. A rain storm such as this may be of high intensity and may last for a few minutes or up to a few hours in duration. Sometimes precipitation comes down as hail and during a cold winter snow may occur at higher elevations. Rainstorms may be preceded by a barometric high over the desert area. In this case, a flow of dry, hot air blows dust from the desert which flows in the direction of the barometric low. In the autumn and spring, when dust storms are most abundant, the hot, dry periods, known locally as "khamsin", may be broken by a heavy rainstorm. Most dust storms are connected with a barometric high over the continent and lows which approach from over the sea. Dust is blown into the lows from the deserts of Sahara and Arabia. When a dust storm blows, visibility may be limited to a few meters and the height of the dust storm may reach 3 to 5 kilometers [8].

The Mediterranean Sea acts as a gigantic temperature regulator, due to the high heat capacity of its water. The further one goes from the sea, the lower the regulatory effect is. As a result of this, the differences between day and night temperature, as well as seasonal temperature, are high. The influence of the Red and Dead Seas and the Persian Gulf, which are enclosed in narrow depressions, is limited to their very close vicinities. Thus, south of the central part of the bow of the Fertile Crescent, the differences between day and night temperatures may reach 15°C, and in some extremes even 20°C. In summer the temperatures may reach 40°C during the day, while during the night they will drop to 25°C; on a winter night they may drop below 0°C, while during the day they may reach 20°C.

The harshness of the high and low temperatures is compensated by the dryness of the weather. During most of the year, the weather is dry. On the one hand, this relieves the stress of the heat, as perspiration evaporates and the human body feels comparatively comfortable when not exposed to direct sun radiation; on the other hand, this causes high evaporation rates from the surface of water bodies and high transpiration rates from vegetation.

Taking into account the scarce and random nature of the precipitation, these phenomena are of crucial importance to the biotic world, as well as to the development of water resources and agriculture, which will be discussed in the following chapters.

To conclude this chapter, a short survey of the soils of the region will be presented. In the more humid area of the region the soil types are a result of climate and rock, while in the desert area they are mainly a result of wind and the flow of water. Types of soils known from the more humid regions are scarce in the desert and are found as pockets or fossil soils.

As previously mentioned, the most typical soil found on the limestone terrains in the more humid areas, is the "terra rosa". On the basalt plateaus, such as that of southern Syria, heavy black soils have developed. The most characteristic soil covers in the desert areas are loess and sand. These soils are eolian, namely, they are brought by winds; the loess is brought as dust from the surrounding deserts and deposited in the more humid areas by rains. The main period of the deposition of the loess was between ca.100,000 years to ca.15,000 years B.P. [9]. Earlier deposits of loess are also known, but in most cases, they are mixed with gravel and bolders and thus belong more to the alluvial rather than to the eolian type of soil.

Sand covers great parts of the deserts of Egypt, Saudi Arabia, and the coastal plain of Israel. Two main sand formations can be discerned, the more extensive being that which is connected with the disintegration of the Nubian sandstone deposits and which forms various types of dunes. Some of the dunes became stabilized, especially during the moist periods which have occurred since they were first formed. The settling of the sand dunes is discerned according to the layer of fines, mainly silts which accumulate on their surface, and the reddish color they obtain due to the decomposition of the iron-containing minerals [10].

The younger layer of sands covering the coastal area of Israel is still mobile. It started its penetration at the end of the Byzantine period, at the beginning of the Moslem conquest, ca. 1500 years B.P. ([16], Chap.1)

In the big valleys, streams and large rivers flowing from the highlands washed large quantities of gravels, sands and silts into the valleys. These materials accumulated to form thick layers of alluvial fans. The upper parts of the fans bordering the walls of the valleys are composed of coarse materials of bolders and gravels, while the lower parts are composed of fine gravel, sands and silts, and the valleys are filled mainly by silts.

In some local topographical depressions, the groundwater table comes very near the surface of the land within the capillary zone. This causes the water to ascend to the surface and evaporate, leaving its salts behind. Flood water, which covers these depressions from time to time, washes some of the salts downward to form a hard layer of gypsum and salts. Thus, the upper soil layers in the depressions are very saline and are underlain by a hard pan which hinders root penetration even for salt-tolerant bushes and trees.

The soils which cover the high plains of the deserts are characterized by a veneer of silts mixed with bolders. The exposed surfaces of the bolders are covered by a brown desert patina. Most of the bolders are flint, which is resistant to erosion and thus protects the fine silt from being carried away by the wind. At a depth of 10-15 cm the silt is highly saline and gypsiferous. These soils are called "reg". Some of these plains are transversed by wide streams which deposited gravel sands and silts. In many cases the fine material is carried away by the wind and a layer of bolders and gravel forms on the surface. This land form is termed "hamada". In most cases, the soils of these areas are less salty than those of the "reg", due to flushing by flood waters. During more humid periods, some of the valleys acted as flood basins, when water coming from the tributaries flowing into the plains does not flow out due to either lack of or too narrow an outlet from these valleys. The water deposits its loads, causing the formation of thick layers of fine silts. The evaporation of the water left behind donates a high quantity of salts. The sodium salts in some cases cause the dispersivity of the clayey materials, making the soil cover impervious. This water evaporates slowly, causing the upper layer to form typical hexagonal buckles [11].

Many of the river-beds in the mountain regions have been terraced by farmers for centuries, presumably since the Upper Bronze Age. The walls of the terraces caused the accumulation of the soils from the hillsides. Today, many of the terrace walls, especially in the desert areas, have been breached and the soils have eroded. In those cases where the soils remain, they are found to be well-drained, good quality soils, lower in salts.

Geology, lithology, and tectonics, together with the climate, have decided the character of the landscapes of the Fertile Crescent. The most southern edge of the Crescent touches the Arabo-Nubian ancient plate, which is built mainly of crystalline and metamorphic rocks. This region, which includes southern Egypt, Sinai, and western Arabia, is extremely arid. The landscape is ragged, with steep mountains and deep gorges (Figs. 9.6, 9.8). The streams flow from to the valley by extensive alluvial fans. The rocks are exposed and are devoid of vegetation. Only in the valleys, along the ephemeral stream beds, may one find stretches of acacia trees. In areas where the water table is near the surface, tamarisk trees will dominate. North of the crystalline shields one finds the plains underlain by Nubian sandstones. The lack of intensive folding, the very low dip of the layers, and the small topographic gradient have given the plains the character of a huge flat plateau with a few protruding table mountains, the tops of which were protected from erosion by a hard layer either of limestone or flint. In many places, the plains are covered by sand dunes.

Only in the erosion cirques of Egypt and the Negev, and along the margins of the Syrian-African rift valleys, where the topography is young, have deep canyons been excavated. In these places, all the sequence of the sandstones is exposed and erosion has sculptured beautiful structures.

Overlying the sandstones, one finds the limestones and chalks of the Mesozoic and Tertiary. These dip gently away from the crystalline core of the continent and also form flat plains (Fig. 2.4). The table mountains which protrude are covered either by a hard limestone or a chert layer. In the river-beds one may find thin alluvial layers and in the wider valleys loess layers and alluvial fans may be found. On the plain, sand dunes brought by the wind from the sandstone exposures are found in Egypt and Arabia. In those regions where the limestone layers are interbedded with chalks or marls, a perched water table may be formed in the more permeable limestone layers. In places where the sequences are cut by a deep ravine, springs may be found. These are marked in most cases by groves of tamarisk trees and reeds. In the lower stretches, date palm groves may thrive.

The terrains built of chalks are much more rounded and flat than the limestone terrains. Due to the fact that the chalks are impermeable, there are frequent occurrences of perched water tables in the alluvial fill of the river-beds. These are also marked by reeds and tamarisk groves. In many cases, shallow wells have been dug into these layers for the exploitation of the water.

Extensive plateaus built of basalts of Tertiary to Quaternary age spread east of the Red Sea and the Jordan Rift Valley. From the plateau rise extinct volcanic cones. The frame and backbone of the Fertile Crescent is built of folded chains of mountains composed mainly of limestones of Mesozoic age (Jurassic and Cretaceous). The topography, especially that of the Taurus and Zagros mountains, is rugged due to their relatively young age, particularly the last upheaval, which took place at the end of the Tertiary and the beginning of the Quaternary. As these mountains rise in the path of the rainstorms coming from the sea, they receive much percipitation. In places where man's hand has not done too much harm, the natural forest of oaks, pines and pistachios can still be found. Yet many of the mountains are bare due to overcutting and in addition to overgrazing. The limestones are very permeable due to extensive dissolution of the rocks by water, and karst caves are abundant. The water is stored in the thick limestone aquifers from which very large springs emerge, give rise to the large rivers which cross the dry plains. The plains form the inner side of the Crescent. They are covered by young Tertiary rocks deposited in shallow seas, lagoons, and continental basins. They are composed of chalks, marls, sands, and gypsum. The topography of these plains is flat and in many instances the streams drain into depressions from which there is no outlet to the sea. Extensive shallow, saline marshes which dry up to form salt and gypsum flats then form. The most extensive of these are those of the Iranian plateau.

In conclusion it can be said that the Fertile Crescent, due to its situation on the fringe of the desert, is also a region where geology, structure, and lithology play a dominant role in deciding the morphology as well as the availability of water resources. They also decide the main patterns of communication, and the main pattern and character of the settlements. It is thus no wonder that a geologist who was fortunate to work in this region not only enjoys the fact that stratigraphy and tectonics are exposed as if to serve as a text book, but is also fascinated by the fact that on the same road that he traveled and near the same spring where he stopped to rest, kings ,prophets, and pilgrims have done the same for thousands of years. It is no wonder, therefore, that, together with his hydrogeological notes, a few additional sentences related to the historical background of the region have been added. These notes and general personal impressions, together with what has been read, in addition to the results of his hydrogeological investigations, have been put together in the present book.

3 The Eyes of Tiamat and the Ark of Noah

And God made the firmament, and divided the waters which were under the firmament from the waters which were above the firmament: and it was so. And God called the firmament Heaven. (Genesis 1:6-7)

Then the lord rested gazing upon her dead body (Tiamat) while he divided the flesh of the body and devised a cunning plan. He split her up like a flat fish into two halves. One half of her he established as a covering for heaven. He fixed a bolt; he stationed watchmen and bade them not to let her waters come forth. ... and over against the Deep he set the dwelling of Nudimmud". (Enuma Elish, Tablet IV 135-142) [4]

And it came to pass after seven days that the waters of the flood were upon the earth. In the six hundredth year of Noah's life, in the second month, the seventeenth day of the month, the same day were all the fountains of the great deep (Tehom Raba) broken up, and the windows of heaven were opened. And the rain was upon the earth forty days and forty nights. (Genesis 7:10-12)

In order to better understand the environmental background of the people who chanted these stories of creation, one has to become better acquainted with the climatic and hydrological regime of Mesopotamia (Fig. 2.5), the country where the first written records of man's history were found [1, 2].

The present climate of the region is semi-arid to arid, with a mean annual precipitation of about 200 mm. The rains fall during the winter, which is rather cold and short (November to March), while summers are long, dry, and hot (temperatures may reach 50°C). The rain storms come from the west and northwest, and are connected with cyclonic low pressure zones travelling from the northern Atlantic and the North Sea through Europe and the Mediterranean. The low quantity of precipitation and the hot, dry weather prohibit agriculture, which is dependent solely on the rains. The livelihood of the people is thus dependent upon irrigation, using water brought from the two rivers, the Euphrates and the Tigris, by means of dykes, canals, and sluices [3]. The water from the rivers is fresh, but evaporation is high, whereas the groundwater is shallow and saline. Thus, if water is applied in excess and soil is not drained well enough to keep the water table below the capillary zone (i.e., below the level at which water can reach the surface due to the force of adhesion) water will reach the surface and evaporate, forming a salt crust on the soil. The high floods of the rivers come from the mountains of Anatolia and Iran between April and June. The reason for the floods occurring so late after winter is that the water is stored as snow on the mountains and as groundwater in the cavernous limestone in the subsurface. The floods are not always a positive factor, as the rivers may overrun their banks and flood the flat surrounding country. This, in a region where the only building material is clay, may be hazardous. Moreover, when the floods and rains happen at the same time, the groundwater table may rise to the surface and the whole plain turn into a huge swamp from which only the tels (the sites of ancient towns which were built each on the ruins of another) protrude.

On the other hand, when insufficient rains and snow fall on the mountains due to a dry year or climatic cycle, less water can be diverted from the rivers, resulting in less food. Thus, for the people in the plains of Mesopotamia, life depends on the balance between the rains over the mountains, the floods in the plains and the subsurface water table, in other words, the water of the sky, in the rivers, and in the subsurface. In the lower stretches of the rivers in the area of the Shat-el-Arab, large expanses of swamps are found and no border can be distinguished between saline sea water, saline groundwater and freshwater from the canals of the rivers (Fig. 3.1). These conditions have persisted more or less since the end of the Last Glacial period, about 10,000 years ago, when the present climatic belts stabilized, although fluctuations of more humid or dryer periods extending over decades or even hundreds of years may have changed this general picture to some extent.

Upon understanding the climatic and hydrological characteristics of Mesopotamia, it is with extreme interest that one reads the ancient texts written on clay tablets approximately four to five thousand years ago during the Sumerian civilization. These texts describe the creation of the world as understood by these people, expressed through analogy to their environment [4, 5]. The freshwaters were represented by the god Apsu, the salt water of the sea and subsurface by the goddess Tiamat. The god Mummu probably represented the mist or even rain. The tablets tell us that Apsu, Tiamat, and Mummu did not like the behavior of the gods who were their descendants, and wanted to destroy them and thereby to destroy the newborn world. This scheme, however, was opposed by Ea-Enki, "He of supreme intelligence, skillful, ingenious, He who knows all things." He put a spell on Apsu, who fell into a deep slumber, then he locked up Mummu, passed a string through his nose and sat holding him by the end of his nose rope.

Evoked by the forces of chaos to avenge her husband, Tiamat mobilized all the forces of destruction. Enlil, in the Sumerian version, or later Marduk, the chief god in the Babylonian Pantheon, combated Tiamat with the help of the winds, killed her, and then cut her body in two. One part he placed above earth to create the sky, while he made sure that the water in it would not escape by setting up locks and appointing guards on the locks. The lower part formed the "water of the deep".

This story, which most probably stems from Sumerian mythology, but was later retold by the Akkadians who introduced archaic Semitic traditions, describes the war between the gods, whether Enlil, Assur or Marduk, protecting their cities, and the forces of destruction symbolized by the powers of the waters of all types. The fact that the evil gods wishing to bring destruction on their descendants were represented by forces of water is an indication, most probably, of the continuous struggle of the societies worshipping the good gods against flooding from the rivers, the upsurge of fresh or salty groundwater, and the encroachment of the saline water from the sea into the shallow delta bordering the Persian Gulf.

Fig. 3.1. Satellite image of the Shat-el-Arab flowing into the Persian Gulf. (Produced by NASA)

This struggle, which started when man first set foot in the Valley of the Two Rivers in prehistoric times, and continues to the present day, was symbolized by the ancients as a war between the protecting and the destructive gods. The latter were also symbolized by serpents and dragons.

The echo of this symbolic fight later saw many variations and has been told and retold by many generations, each claiming their protector to be the hero who killed and put the spell on the forces of evil. The struggle between the hero-god patron and protector of human society, and the forces of destruction, symbolized by Tiamat, a Dragon, or the Tanin, knew many incarnations; St. George killing the dragon is just one of them.

The question is: Where did this basic idea of the water as a primeval force of destruction tamed by a divine constructive force originate? Was it in the fertile valleys of Mesopotamia or Egypt, or in the plains and valleys of Canaan, Anatolia or Iran? Surveying the hydrological conditions of Mesopotamia, one arrives at the conclusion that indeed the origin of this theme of war between the powers was in the plain of Mesopotamia, most probably with the Sumerian people. The Akkadians, coming with their stories of creation into a new environment and finding a better answer to their questions in the local stories, adopted them with some modifications. In Mesopotamia the water, when uncontrolled as either floods or saline water encroaching the fields, indeed meant destruction. This brought about the organization of a social system, the tasks of which were the building of canals, embankments, and drainage channels, and their constant maintenance. This organized society was governed by a king, who received his power from the divine protector of the city, which afterwards expanded to become a state.

The story of Marduk overpowering Apsu, Tiamat and Mummu, as told in the "Enuma Elish", the Babylonian "Genesis", was recited at the festivities of the New Year in Babylon (Bab = gate, Ilim = gods). During these festivities, that were celebrated at the beginning of spring, namely the first days of Nissan, the first month of the Babylonian calendar, the king went through a ceremony of dethronement followed by re-coronation. These festivities of the New Year were the time to repent and request for mercy in order to protect man from the destructive powers of nature. These New Year festivities were celebrated at the end of the winter, namely, when spring brings with it the flourishing of vegetation, of the grasses in the pasture lands, the blossom of the trees, and the birth of young.

As already mentioned, the Babylonian tradition of the victory of Marduk on Tiamat is known to be based on Sumerian mythology, the literate history of which began about 5000 years ago. It is quite obvious that the traditions came down from earlier times when stories were passed verbally from one generation to the next. The mentioning of water, as well as the many symbols connected with it, points to the fact that the traditions grew in a country which was endangered by, as well as benefiting from water. As the Sumerian civilization had had its roots in the Mesopotamian plain since prehistoric times, it is logical to look for the roots of these mythological traditions in the interaction between the natural environment and man in this area. In Sumerian mythology, the fight between the benevolent and the evil gods also contains the subject of primeval waters ruled by the evil god. In one such story, Kur, the god of the netherworld, who controls the Primeval Waters, is involved in a foul deed against the sky goddess, Ereshkigal. Enki, the water god, sets out in a boat to attack Kur, who fights back by sending the primeval water to sink Enki's boat. The tablet does not tell the end of the battle, but from other sources it is known that Enki, the good god of water, triumphed.

This tablet starts with the following lines:

> After heaven had been moved away from earth
> After earth had been separated from heaven
> After the name of man had been fixed... [2]

Thus once again one finds the tradition of creation by the separation of Heaven from Earth connected with the story of a fight against the Primeval Waters.

The story of Tiamat and Marduk, and also that of God, dividing the land into the sky or upper water and sea or lower water, can thus be seen to have developed not just from one ancient story, but from an accumulative process of experiences starting in prehistoric times when man had just begun to penetrate the plains of the Euphrates and the Tigris, and continuing through the period when history started to take shape.

The picture which evolved before the eyes of the intelligent Homo when he first came into the plain of the Euphrates and Tigris, was that of a land coming out of chaos into some kind of pattern; a marsh of saline and fresh water covered by mists which does not enable the boundary between sky and land or water to be seen. (Fig. 3.1).

Another possibility is that the fertile land emerging from the domination of the waters was a land of clash between two races of people: on the one hand, the people coming down from the highlands where either overpopulation or reduction in precipitation forced part of the population to immigrate, and on the other hand, people coming from the fertile plains of the Levant, which was slowly turning into deserts. The story of the fight for life or death between Cain, the land tiller, and Abel, the pastor, may be an echo of this clash, which ended by these lands adopting the material culture of Cain, while they also absorbed the beliefs and religions of the pastoral people.

Thus, the story of the creation of the land out of the primeval water and chaos was passed on from generation to generation. Returning to these stories of creation once the picture of the natural environment has become clear, one cannot but admire the verbal power of the ancients in conveying to their descendants the natural processes which they experienced.

> When in the height heaven was not named,
> And the earth beneath did not yet bear a name,
> And the primeval Apsu, who begat them,
> And chaos, Tiamat, the mother of them both,-
> Their waters were mingled together,

> And no field was formed, no marsh was to be seen;
> When the gods none had been called into being,
> And none bore a name, and no destinies were ordained;
> (Enuma Elish, The First Tablet from the Seven Tablets of Creation.)[4]

Granted that these myths deliver to us the beliefs and fears of the settlers of the plains of Mesopotamia, as well as their explanations of the wonders of nature they experienced, the question remains, can we correlate these stories with specific historical events, or are the myths just some kind of a mixture of many events that have lost any traceable connection with what really happened?

One of these stories is that of the Deluge, which is also one of the most prevailing mythologies of the peoples in the Fertile Crescent. In 1827 the members of The Society Of Biblical Archeology in London were astonished to hear from the assyriologist Mr. George Smith that he had deciphered this story on an ancient clay tablet on which it was written in the Akkadian language in cuneiform writing [6]. Since then, older versions of this story have been deciphered and today it is quite obvious that it stems from Sumerian origins (see General Bibliography). Most probably, it passed from one civilization to another, each explaining it in its own way according to its set of beliefs and faiths.

During the years 1927 to 1929, the archeologist Sir Leonard Woolley and his team were excavating in Lower Mesopotamia in the graveyard of the city of Ur (Fig. 2.5) from which "Terah took Abraham his son, and Lot the son of Haran his son's son, and Sarai his daughter-in-law, his son Abram's wife, and they went forth with them from Ur of the Chaldees, to go into the Land of Canaan and they came into Haran and dwelt there" (Genesis 11:31). The archeologists were excavating a shaft in a thick layer of rubbish which for centuries had been dumped outside the walls of this ancient city. The shafts went deeper and suddenly the character of the soil changed. Instead of the stratified pottery and rubbish, they were in a layer of "perfectly clean clay uniform throughout, the texture of which showed that it had been laid by water". The workers thought that they had come to the river silt. Woolley, however, felt sure that they were still high above the level of the marsh. After calculating the measurements, he sent the men back to work and, indeed, after excavating through 2 to 3 meters, they again found themselves in "layers of rubbish full of stone implements, flint cores from which the implements had been flaked off, and pottery". Sir Leonard concluded that he had found evidence of the biblical Deluge, "which was not universal, but a local disaster confined to the lower valley of the Tigris and Euphrates affecting an area perhaps 400 miles to 100 miles across" [7]. This conclusion however, did not convince the other archeologists who were working in Mesopotamia. They argued that at Eridu, only 15 miles from Ur and lying somewhat lower, no trace whatsoever of a flood was found. The sterile clay layers found at other sites, for instance Kish, Uruk, and Lagash, belonged to other periods such as the Early Dynastic, which were later than those found in Ur of the

Chaldees. The conclusion of most archeologists was that archeological excavations in Mesopotamia have afforded no evidence of a regional cataclysmic Deluge.

This conclusion does not conform, however, with the fact that the story of the Deluge is found in all the sagas of the ancient civilizations of Mesopotamia, Babylon, and Assyria, not to speak of the story of the Bible. It even goes back to the Sumerians, who narrated it in a text found at Nippur, dated about 3700 years ago. The question remains whether the story conceals an echo of a catastrophe which happened to the ancient civilizations of Mesopotamia or the Levant, or whether it is a pure myth, namely, a story fabricated on a small incident in a few towns or a group of villages in prehistoric times which, through the millennia, was given more and more of an epic significance by succeeding generations, who added their own adventures to the original story. Their experience of Mesopotamia as a country in which floods occurred from time to time, causing local damages, gave credibility to the story.

The question whether there was a Deluge or not has interested, and still interests, many people for various reasons. On one side stand the Biblists, who are searching for evidence which will convince them of what they are already convinced of, namely that every word in the Bible is true. On the other side, stand many scientists, who would like to prove that the Bible, as any other mythology, tells more about the subconscious of the primitive folk who invented it than about what happened in historical or even prehistoric times.

As was mentioned in the Introduction to this study, the autl r suggests another approach which he believes is more objective. He suggests regarding the biblical stories as an echo and a possible source of evidence of some distant traumatic experience through which the ancestors of our civilization may have gone. These traumas were a result of either natural or anthropogenic causes. In many cases the cause may have been a complex one, as, for example, a natural event such as a climatic change bringing a wave of cold which drove hordes of people from one region to another, spreading war and destruction.

There are two main reasons that led the author to adopt this approach. In the first place, the Bible, which forms the foundation of his Jewish cultural heritage, portrays a gradual evolution from mythology to history. As episodes narrated during later periods were found to have their archeological proofs, there is no reason not to assume that stories from earlier periods had some factual roots, too, at their source. In the second place, it is quite natural and scientific to examine every piece of evidence upon which one can put his hands, whether archeological, geological or anthropological. No one would argue that the Bible can be regarded as a document in which, at least, an echo of the experience of former generations can be found. The condition is that one should not take this evidence as words coming directly from the mouth of the Almighty and thus leaving no place for any skepticism and scientific enquiry.

Returning to the story of the Deluge, the two best-known tales are the one told in the Bible about Noah and the Babylonian story of Gilgamesh [8,10]. The moral is entirely different in each of the two stories. In the Bible, the destruction of mankind by the Deluge came as punishment for their evils and Noah was saved because of his righteousness. In the Tale of Gilgamesh the destruction came because the gods quarrelled, and Utnapishtim was saved and became immortal only because the gods favored him. An earlier yet less known story of the Deluge is found in the story of Atra-Hasis [9)]. It is also a Babylonian epic story but it bears more resemblance to the Sumerian version of a story on the flood. (In the Sumerian story, King Ziusudra is the survivor of the flood). In Atra-Hasis the flood comes as a punishment to mankind who made too much noise and disturbed the sleep of Enlil.

The reexamination of these stories on the background of our understanding of the climatology and hydrology of Mesopotamia leads directly to the interesting observation that the hydrological phenomena described in all the stories of the Deluge are very similar, but at the same time are different from what one would expect in the framework of present climatic circumstances. All of these stories describe the flood as a result of heavy rains and the emergence of water from the subsurface. (On the clay tablet, with the Sumerian story of King Ziusudra, the beginning of the narration is destroyed). For people living in a country where the main source of water for everything was the big rivers, while rain was normally a marginal phenomenon, this is quite strange. Did the catastrophe happen when these people lived in another environment, or was it the drastic change in the natural circumstances that left its strong impression on the people, to be remembered for many generations? It seems quite probable that if such a flood, caused by heavy rains, had happened in a land where such an event happens rather frequently, such as the monsoonal countries, it would not have left such a deep impression.

The explanation suggested in the present work is that indeed a true yet unusual environmental catastrophe lies at the roots of the story of the Deluge. In other words, the story was composed of a few different layers, at the basis of which lies a distant memory of events characterized by a change in the climatic regime. This entailed rains which flooded the plains and in the same time caused a tremendous increase in the flow of the mountain springs feeding the rivers, due to a rise of the groundwater table in the limestone aquifers of the Taurus and Zagros mountains. Only the people who were clever enough to embark on a floating device, either an ark or boat, and steer to the mountains for refuge were saved.

When did this happen? Hallo and Simpson from Yale University, in their book The Ancient Near East [11], concluded that the great Deluge of Mesopotamian sources was somewhat more localized than the mythological picture, but it was nonetheless a historical event associated with a specific point in time. They suggest that it happened around 2900 B.C. (or 4900 B.P. - Before Present). They base this suggestion on the fluviatile layers which were found in some of the archeological excavations in Shuruppak. There the flood deposit "while more modest, is definitely of a fluvial character, moreover it intervenes precisely between the Jemdat Nasr and Early Dynastic levels, that is at the very point when the relative chronology of native sources places the flood". Examination of Fig. 1.3 shows that indeed about 5000 years B.P (3000 B.C.) the ^{18}O and ^{13}C isotopes in the sediments of the caves and Sea of Galilee are depleted, which speaks for a cold and thus humid climate in the Fertile Crescent.

Yet, the findings of Woolley suggest that this was not the only period during which such floods have occurred. There were many more floods in these regions, so that the stories of floods piled up in the collective memory of the people of Mesopotamia. The author came to the same conclusion after examining the results of a paleoclimatic investigation carried out on the sediments of Lake Van in Turkey [12,13]. This examination has shown that such changes did occur (Figs. 1.3, 3.2). As the water of this lake comes from the same mountains which supply the Euphrates and Tigris and as the lake is closed, the water and sediments which flow into it remain and accumulate. Core-drilling and sampling of the sediments found at the bottom of the lake (samples were taken out in a barrel-like pipe and their contents investigated), showed that materials brought by the rivers flowing into Lake Van were deposited as thin varves, each built of a white layer which was deposited during summer, when evaporation was high. A black layer was deposited during winter when clay and organic materials were brought by the floods. The varves were counted and the age of each layer calculated, as well as the rate of deposition, namely, the thickness of the varves for each year. Though it is argued [13] that the number of varves does not represent the true age, due to differences in the mode of deposition, no one argues against the fact that they portray changes in the mode of deposition due to climatic changes. The pollen of flowers, trees, and herbs were preserved in the sediments and the counting and ratios of various species could show the type of vegetation which grew in the drainage basin.

The counting of the varves in Lake Van has shown that the deposition of the layers drilled by the wells started at least about 14,000 years B.P., at the end of the Last Glacial period which was also the transition from the Pleistocene to the Holocene periods. The layers at the bottom of the deepest drill hole showed that during glacial times the climate was cold, with many floods with a high rate of sedimentation filling the basin. Between 10,000 and 9000 B.P. the climate turned warm and dry. This caused a fall in the level of the lake and an increase in the salinity of the water. The pollen found in the sediments show a high percentage of herbs and a small percentage of trees. Around 6500 B.P. there was a dramatic change in the environment. This is reflected in a rapid rise in the level of the lake a decrease in the salinity, and a marked increase in the percentage of pollen of trees, especially oaks, to that of herbs. From 6500 to 3400 B.P. only small fluctuations can be seen in the climate as recorded in the sediments of the lake. These fluctuations, give the long term climatological background which persisted during the evolution of the civilization of Mesopotamia, For the purpose of distinguishing events of shorter periods one has to examine also the diagram of the rates of sedimentation in the lake (Fig. 3.2). These are proportional to the sediment load of the rivers flowing into the lake, which might be proportional to the volumes of water brought by the rivers.

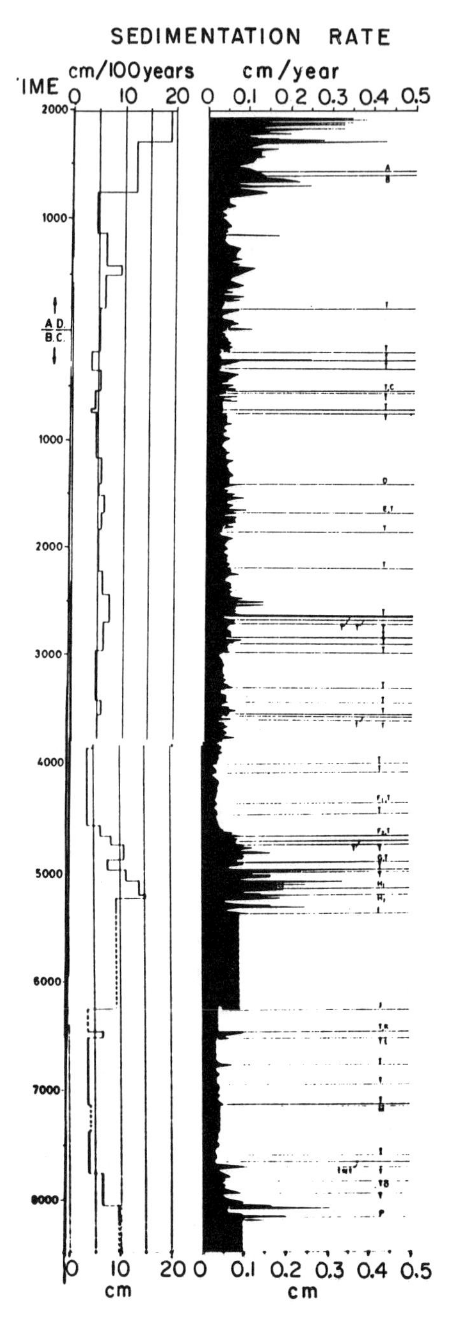

Fig 3.2. Sedimentation rates in Lake Van, Turkey [12]

Floods in these rivers are an indication of floods in the Euphrates and Tigris. The information one obtains from this diagram is that, although floods of a high discharge are abundant throughout the history of the lake, there are periods in which these floods came in high frequencies. Such periods are most probably those which may have caused the flooding of the plains, as they meant periods of higher precipitation rates and higher flow in the river-beds, which together have an accumulative influence.

The sedimentation rates per 100 years show peak values for the period at about 7000 B.P., and extend to the period of 6300 B.P., namely, the period when the lake began to fill up. Another period of rather high values is that from 4900 to about 4300 B.P., which fits the dates suggested by Simpson and Hallo. In Lake Zeribar,in Iran, the pollen analysis shows that, at about 5000 B.P., there is a maximum in the ratio of pollen of oak, which denotes a more humid climate [13]. Taking into account the inaccuracy in ^{14}C dating, this may also agree with Hallo and Simpson's dating.

Another record from which climatic changes can be inferred is the acidity and ^{18}O content of the ice cores of Greenland [14]. This information is, however, global, and does not indicate how a certain change influenced a local system as do the cores from Lake Van to Mesopotamia. Moreover, lately the exact dates of the various events have been questioned [15].

From the acidity of the ice cores one can learn mainly of strong volcanic explosions, containing gases rich in sulfur. This, as mentioned, was correlated with the reduction in ^{18}O content of the ice which is indicative of colder climates. In the opinion of the scientists who investigated these cores, the volcanic explosions threw a large amount of ash into the atmosphere, which caused a reduction in the solar radiation and thus lower temperatures. In these cores, one can find supporting evidence for the triggering of a colder period at ca. 6400 years B.P. and from ca. 5200 to 4700 B.P. and thus supporting evidence for the prehistoric flood envisaged by Hallo and Simpson. (Though the exact chronological correlation may change in the order of magnitude of 100 to 200 years).

New evidence, which has already been mentioned, is from the core samples taken from the Sea of Galilee in Israel. As explained, the rain storms come to Mesopotamia mainly from the west, also crossing Israel. Thus information from this land is relevant to Mesopotamia. A study carried out on the environmental isotopes as well as the pollen of this core [14] showed that, after 5000 B.P., there was a reduction in the ^{18}O content of the sediments of the lake (Fig. 1.3), which indicates a cooler, and thus most probably more humid period. Again considering the inaccuracy in the dating , this cooling may coincide with the time suggested by Hallo and Simpson.

In conclusion, it is suggested to consider the mythological stories of the flood of Mesopotamia and the story of the Biblical Deluge, as an echo of pre- and proto-historical real events which occurred while and after man settled in the fertile valley of the two rivers. The main impact on his heritage was by the events which occurred after man learnt to harness the water of the big rivers for his benefit and learnt about the sources of these rivers from big caverns in the mountains, as well as of the existence of underground waters in the plains. Then suddenly, these waters went out of control and brought destruction to all his ordered and organized world, his whole traditional world view got disturbed and he looked for an explanation for this disturbance.

In the pagan world the forces of nature were represented by a multitude of deities ruled by primeval forces of chaos. Thus such a catastrophe was interpreted as a disturbance in the ordered life of his gods, by a war or just by a quarrel between them, in other cases as man's failure to meet their desires for food and presents. To the Hebrews, justice commanded by their abstract Lord was the foundation of order in the world and such a catastrophe was interpreted as a punishment for the corrupt ways of mankind.

In addition to these conclusions regarding the correlation between the stories of the ancients and climatic events, it is the opinion of the author that through a careful study of the biblical text and its comparison with the ancient Babylonian texts, and keeping in mind what is known of the hydrological environment of Mesopotamia, one can learn quite a lot about the world view of these ancient societies. One can see that indeed many of the archaic traditions originated in Mesopotamia and had been transferred from one environment to the other, possibly verbally, and that the Hebrew people who afterwards put them into writing did not know the exact meaning of all the words in the archaic Akkadian or even Sumerian text. For example, in the Akkadian version of Atra-Hasis, which might be the nearest to the Sumerian original, the god orders Atra-Hasis to use his hut built of canes ("kanim" in Akkadian and Hebrew) and pitch it (Put "cofer"). The biblical text, although using words identical to the Akkadian for pitching, changed the type of wood to "gofer" which is very similar to "cofer", namely pitch, and instead of kanim (canes) uses the word "kinim" for which the nearest Hebrew meaning is birds' nests interpreted traditionally as "divisions" or "small rooms."

Another interesting observation is related to the term "Mayanot Tehom Raba", translated as "the fountains of the deep" (see the citation from Genesis at the beginning of this chapter). In the opinion of the present author this term is an indirect reference to the archaic goddess of the deeps, Tiamat, whom Marduk had cut into two. From the upper part he formed the sky, or the upper water, and the lower part became the sea, or the lower water. In the story of the Deluge and, previously, in the biblical story of the creation, the term "Tehom" (Genesis 1:2) already had lost its direct association with the fight between Marduk and Tiamat, and Tehom is used as a noun to describe the primeval and lower deep water. But although Tiamat and Marduk, were forgotten, still the same conceptual model of two bodies of water reoccurs in the biblical description of the Deluge. The water from two sources: from the "windows of heaven" and from the "fountains of the deep".

The survival of the term Tehom-Raba as a body of water even after Tiamat was destroyed and no longer considered a deity, in the author's opinion, throws light onto the archaic source of the term ayin or "eye", which in Hebrew and Arabic means fountain as well as spring. (It is interesting to note that the Persian [an Indo-Aryan language] term for spring is "cheshme", which is related to "chash", which means eye.). It seems that the archaic pictorial concept, and thus the explanation as to the appearance of springs, was that they were the eyes of Tiamat. The word ma'ayan might thus be an ancient complex form of the words mei'ayin, namely "water of the eyes" (of Tiamat) or m'ayin "from the eyes." The association of springs with the eyes of Tiamat seems quite natural when one takes into consideration that the two large rivers, the Tigris and the Euphrates, emerge from large springs, most of them karst springs (or limestone solution outlets) along the foothills of the Zagros mountains on the border between Iran, Turkey and Iraq. These springs in many instances form large circular pools of water out of which a stream flows (Fig. 3.3). In many cases the spring emerge from big caverns. In the British Museum one can see on an Assyrian stele a pictorial description of Shalmaneser III's visit to such a cave at the source of the Tigris. The relief shows cattle sacrificed at the source, while workmen are busy cutting commemorative reliefs on the rock [17]. When severe rain storms occur on the mountains, the springs respond with a strong outflow, causing the rivers to swell and overflow the plains. Thus, the safety of the people living on the plains was connected with the harnessing of these springs and rivers, i.e., the killing of Tiamat and closing her outlets. Therefore it is suggested that the phrase "Mei Tehom" or "fountains of the deep" be considered as a remnant of the archaic story of the Deluge which referred to the water as coming from the eyes of the goddess Tiamat and the water coming from the holes in her upper part, which formed the skies. See also Cassuto [18] for the correlation with the Babylonian text of the sealing up of the part of the body which formed the sky to prevent the water from draining out.

It is also interesting to note the close connection found in the Bible between the slaughtering of the snake (Nahash Bariah), or the dragon Tanin and the great deep Tehom. "Art thou not it that hath cut Rahab and wounded the dragon? Art thou not it which hath dried the sea, the 'waters of the great deep'?" (Isaiah 51:10). Such legends in their most archaic form are found in the Sumerian mythological tales of creation and repeated in various forms in Akkadian, Babylonian, and other archaic mythologies.

Another interesting detail of the story of the Deluge is the landing of The Ark on the mountains of Ararat, These mountains are located on the border between Turkey, Iran and Armenia, far away from the plains of Mesopotamia. Whether these are exactly the ones that the ancient narrator had in mind is not sure. Yet these mountains are very impressive when looked upon from the south (Fig. 3.4). They are extinct volcanoes, which consist of two peaks. The higher one, seen in the figure, reaches an altitude of 5,205 meters. Its top is permanently covered by snow (picture taken in August) and glaciers occur in some valleys. One can thus conclude that during colder periods the glaciers moved downwards, covering everything in their path, and also causing the deposition of typical glacial sediments. When they retreated, at the end of the glacial periods, the sediments were eroded, exposing buried trees, like the ones found in such sediments in the Alps. These were, most probably, interpreted by the ancients as remnants of the Ark.

Fig. 3.3. A karst spring flowing from the limestone rocks, on the border between Iran and Iraq (Photo by the author)

Fig. 3.4. Mount Ararat,Turkey (Photo by the author)

4 Adam and Eve Depart, Enter Cain and Abel

> In sorrow shalt thou eat of it all the days of thy life. ...In the sweat of the brow shalt thou eat bread. (Genesis 3:17,19).

Although the oldest written documents come from Sumer, the archeological investigations show that civilized societies existed in the Fertile Crescent before man settled in the Mesopotamian plain. Man was in the Fertile Crescent already at the dawn of the history of mankind; as a matter of fact, he was there even before he could be called man. Some type of a mythology might even have developed which man brought with him to Mesopotamia as a verbal tradition before he knew how to put it in writing. In the following chapters the more detailed geology and hydrology of the areas from which such societies could have come will be examined and possible influences of the hydro-environment on the origins of the mythologies will be examined.

Prehistoric man's choice to settle in this region, which is semi-arid due to its position on the margin of the arid global belts, is an obvious indication that this border was not static and that the semi-arid zone was more humid during certain periods. Indeed, one can find signs of colonies of prehistoric man in the middle of the desert near sites which were once either swamps, lakes or springs, and dried up when arid conditions prevailed. At these locations were found flint tools, bones, and teeth of animals killed by prehistoric man. Sometimes even human bones have been found. Thus, throughout most of his history, prehistoric man's mass penetration into the desert followed a climatic change when more humid conditions prevailed in these regions.

The earliest flint tools were found in the Jordan Valley south of the Sea of Galilee [1]. At this location, artifacts which were produced by the edging of a pebble were found in abundance, together with bones and teeth of animals such as hippopotamuses, crocodiles and elephants, which today live near tropical swamps. The assemblage of the implements resembles very much that found in the upper beds of Olduvai in Eastern Africa where the most ancient remains of a prehistoric hominid were also found. The age of the layers containing these tools in Africa, dated according to the ratio of potassium to argon, was found to be about 1.2 million years. Although there is no possibility of exactly dating the layers in the Jordan Valley due to the lack of an interlayered volcanic rock on which the potassium-argon method used in Africa can be applied, the correlation allows us to claim that hominids began living around the swamps of the Jordan Valley about 1.2 million years ago [1]. A culture a little younger than the pebble tool culture found in the Jordan Valley, called the Abbevilian (according to the site in France where it was first found), was found in the Galilee [2]. Flint tools of the "Hand Axe" culture (called Acheulian, also according to the name of the site in France where these tools were first located) of the Lower Paleolithic were also found in the arid parts of the Fertile Crescent [3]. In many cases, these contained also bones, teeth, and tusks of elephants, hippopotamuses, and crocodiles. This indicates that during the period which extended from about 700,000 years B.P. to about 100,000 years ago, the country went through a few spells of more humid climate, and hunting groups could live around lakes or swamps in areas which are today very arid.

The time of the Mousterian culture period, (Middle Paleolithic) that followed was most probably also humid and cold. Layers of this period, together with the skeletons of an evolved type of Neanderthal hominid, were found in caves on Mount Carmel and in the Galilee [4]. This was the period of the Last Glacial. It continued from ca. 80,000 B.P. to about 14,000 B.P. The humid conditions which characterized this period in the Fertile Crescent terminated at about 10,000 years B.P. The climatic conditions during these periods need special elaboration as they serve as the key for understanding the climatic changes during historical times

The decoding of the paleoclimatic picture started with the observation that the age of the fossil groundwater under the Sinai and Negev is that of the Last Glacial period and that the isotope content of this water (^{18}O and deuterium ^{2}H) is different from that of the water of the present. Later, another investigation showed that rain storms which come to the Negev over northern Africa have an isotopic composition similar to that of the fossil water [5]. This type of rain storms was preceded in many instances by dust storms. This coincidence led the author to suggest the conceptual climatic model, already mentioned, which claimed that during glacial times the desert and the rain storm belts shifted to the south, causing heavy dust storms, the dust being precipitated by the rain . This in the first place explained the formation of the loess layers peculiar to the period of the Last Glacial and the desert borders [6]. It also explained the encroachment of sands over the loess at the end of the Pleistocene and beginning of the Holocene, as the shift of the belts northward during this period brought Ethiopia under the monsoonal regime, which increased the supply of floods and sediment load through the Nile to the sea and from there to the coasts of Sinai and Israel [7].

The end of the Mousterian cultural period also witnessed the disappearance of Neanderthal man. Did he evolve into a more successful hominid or a more fit descendant, namely Homo sapiens, or was he exterminated by another race, which evolved somewhere else, then invaded the region once dominated by Neanderthal man? To this question there exists no answer yet. What we know for sure is that the

Upper Paleolithic man was not different in his skeletal characteristics from any human type of the present, although slightly different racial characteristics began to evolve. Evolution since this period has been mainly cultural. At about 14,000 B.P. the process of deglaciation took place. The climate changed in the Middle East and became drier and warmer. At about 10,000 years B.P. a new period, the Holocene started, the climate was more or less the same as today. At about 9000 years B.P. a major change took place in the way of life of many human communities. Man started to cultivate the land. How and where this happened will be described in the following paragraphs.

In humid countries, the availability of water is taken for granted. It is one of the basic items of which nature seems to be composed. However, the sun and the warming fire are things which are not guaranteed daily and thus have to be cared for and thought of perpetually. In arid and semi-arid countries, on the other hand, water is the essence of life which is not guaranteed. Its appearance is irregular and random and when it fails to appear famine and death are the consequences. When the same irregularity and randomness work on the other extreme, it brings an overabundance of this resource leading to destruction and chaos. Thus, it is not surprising that the primeval deities (such as Tiamat, Apsu, and Mummu) of the societies living in the Mesopotamian plains that we have already described were those symbolizing the various forces of water. The victory of another force, either Marduk, Bell, or Asur, on the other hand, symbolized a decline in man's fear of these forces of nature after he had succeeded in regulating them. This he succeeded in doing also by organizing human society in such a way as to enable it to tame rivers and build canals and irrigation networks.

Thus it is quite possible that the shift from a hydraulic Pantheon to warriors and agricultural gods was due to the submission of the aborigine population that worshipped the powers of the local rivers and their floods by people who brought with them their own gods from another environment together with a different organizational system, which later enabled them to undertake construction of large projects.

Support for this idea of a new regime and "a new order" can be seen in the fact that Tiamat, Apsu, and their Pantheon were always described as destructive forces and as opposing order. As a matter of fact, they are provoked to kill the gods and thus cause the war (at the end of which they were destroyed) because of the "new" ways of the gods who want to install "a new regime" and thus disturb Tiamat's and Apsu's sleep.

From the agricultural point of view, this can be interpreted as the opposition of the forces of nature represented by the gods of the people living in peace with them, to the interference of patron deities of newcomers who want to regulate the water resources and benefit from them by using them for irrigation.

Thus, one can assume that the archaic societies in Mesopotamia were not agricultural. This assumption is supported by evolutionary, technological considerations, because in these plains too great a human effort is needed on an extensive geographical scale in order to tame and then benefit from the big rivers.

Any small-scale endeavor is doomed from the beginning to destruction by the next year's floods. Thus, just by analyzing the message hidden in the ancient myths, one becomes suspicious that the societies who wrote them and built the big water projects were not the aborigines of these plains but came from somewhere else. Moreover, it is more logical to assume that the first river regulating projects were initially started in the small mountain valleys by diverting and controlling the flow of water coming from small rivers fed by springs, the perennial flows of which supplied drinking water as well as water for irrigation. Only after learning how to control and benefit from a small but regular and manageable flow, can man venture to measure himself against the giants, the flow of which changes drastically from one season to another.

Where could such an evolution have taken place and did these people leave any record of the first steps of their evolution? An answer to this question is found in Jericho in the Jordan Valley where the British archeologist, Kathleen M. Kenyon, excavated the remains of an urban community which was found to have practiced agriculture methods at the early dawn of civilization [8].

Jericho is situated in the Jordan Rift Valley about 10 km north of the present northern edge of the Dead Sea (Fig. 4.1). It can claim the title of the lowest town in the world as it is located at about 300 meters below sea level. It can also claim the title of one of the oldest known towns in history, if not the oldest. The name Jericho comes from "Yareakh", which means moon (as well as month) in some Semitic languages. Was Jericho a place of cult to the Moon Goddess or God? Was the place already called by this name in prehistoric times? This we do not know, but what we do know is that the earliest traces of an urban agricultural community were found in this place. As it is situated in the arid valley of the Jordan river and its perennial water supply is dependent on a large spring, one can claim that indeed dependence on irrigated agriculture may have started here.

Jericho is an oasis, the life of which depends on the spring called in Arabic "Ein-el-Sultan" (The Spring of the Sultan) and in Hebrew "Ma'ayan Elisha".(The Spring of Elisa). This spring is fed from the rains falling and infiltrating into the permeable limestone mountains of Judea and Samaria (Fig. 4.2). Its flow is about 10 million cubic meters per year. The flow is more or less steady throughout the seasons although some minor fluctuations can be observed. The multi-annual fluctuations are also rather small. What is the reason for this regularity in such an arid environment? The answer is subsurface storage, which is the characteristic feature of groundwater. In other words, the physical character of the permeable subsurface enables the accumulation of the varying quantities of rainwater recharging the groundwater aquifers, its storage in the subsurface and its gradual release. This is the principle which enabled man to build a stable culture based on irrigation in this arid land.

At this point the principles of groundwater storage should be explained in more detail and how these enabled the beginning of man's earliest attempts to become independent of seasonal fluctuations of the climate. (For more details the reader is referred to Appendix I).

Fig. 4.1. A NASA satellite image of Central Israel

After the rain reaches the surface of the earth, each drop of water is subjected to the force of several energy fields. A thermodynamic one may cause it to evaporate. The roots of plants absorb the water by osmotic suction and transpire it. Some of the water which does not evapotranspire flows on the surface as runoff, which will join other streams to flow as rivers to a lake or the sea. Some finds its way downward through the pores and fractures of the rocks. This water is pulled by the gravitational force of the earth, and when it reaches an impermeable layer it accumulates on it while saturating all the pores and openings in the permeable aquiferous rocks overlying the impermeable layer. The water in the porous rocks

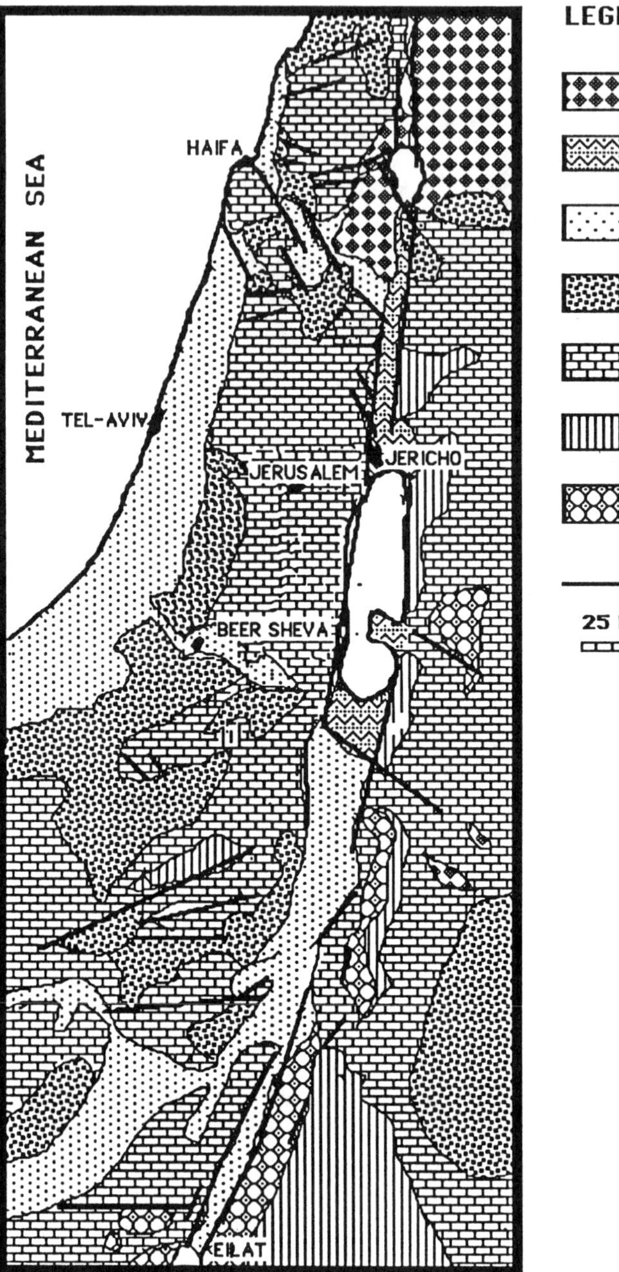

Fig.4.2. General geological map of Israel

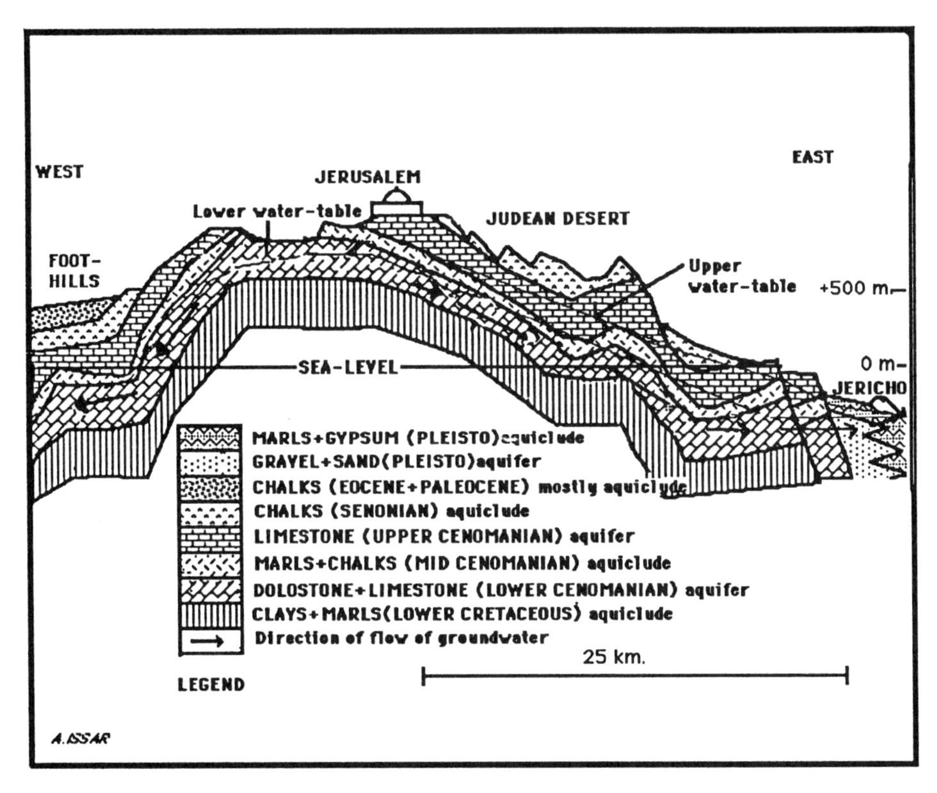

Fig. 4.3. Hydrogeological cross section, Central Israel

(aquifers) flows semi-horizontally toward the lowermost outlet. The plain between the saturated and non-saturated zone is called the groundwater table. In the places where the groundwater meets the surface, a spring emerges. The flow of water from the spring is a function of the amount of water which was recharged into the subsurface by the rains, but the volume of the water stored regulates the outflow; thus, the bigger this storage, the higher the mode of regulation of the outflow. The spring of Jericho has its recharge area on the permeable limestone layers of Middle Cretaceous age in the mountains of Judea and Samaria. There, the waters from the winter rains percolate into the subsurface. They form a few small springs perched on impermeable layers which emerge in the mountainous region and on which the communities in these areas depended. The rest infiltrates deep into the subsurface and flows partly to the west toward the Coastal Plain of the Mediterranean Sea and partly toward the rift valley in the lowest part of which is the Dead Sea. This water emerges as springs at a few points favored by geological conditions. One of the outlets is near the site of Jericho and the tremendous volume of saturated rock in the subsurface of the eastern watershed of the hills of Judea and Samaria guarantees the steady flow of the water, (Fig. 4.3).

With this background information, one can go back about 10,000 years and reconstruct the early stages of irrigated agriculture. This was the time when the influence of the Last Glacier on the climate of the globe dwindled away and with it the more humid phase over the Levant. The big lake which filled the rift valley from the Dead Sea to the Sea of Galilee during most of the period of the Last Glacial receded. Springs which emerged along the high shores of the lake, at an altitude of about 200 meters below MSL, migrated downward after the receding lake shore. At the same time the climate became more arid. The Epipaleolithic hunter of the Natufian culture, who dwelt on the shores of this lake and was a hunter, fisherman, and grain gatherer, found conditions of life more and more difficult. It took him about 1000 years to adapt to the new climatic conditions. This adaptation was not just a matter of inventing a new or more sophisticated flint tool. This time the change was in his whole way of life. Man found a way to improve on the natural system of production, by the process of seeding and growing wild plants.

This throws light on the finding of remains of grains of domesticated two row hulled barley and wheat in the layers of Jericho, dated by ^{14}C to be ca. 9500 years old. The layers belong to the Pre-Pottery Neolithic period when man could not yet bake the most primitive clay vessels but could sow his field, reap the grain, grind it into flour on a specially manufactured grinding stone, and produce dough and maybe even bread.

The climate of the Jordan Valley in the Post-Glacial period was most probably more or less of the same character as today, namely about 120 mm of rain with an annual evaporation potential of about 3000 mm. But even if the climate was more humid and precipitation was twice that of today, no agriculture could have been maintained without irrigation. The proximity of the town to the spring and the finding of the domesticated grains favor the conclusion that people from the Pre-Pottery period did indeed know how to till their land and control the flow of water, diverted from the spring, in such a way that the soil in which the grains were sown would not be flushed away or over-soaked. As the soils of the Jordan Valley contain a high percentage of salt and gypsum remaining from the brackish water of the Lisan Lake which covered the entire area during the Last Glacial period, and as the high ratio of evaporation causes salinization even of ordinary soils, it seems quite probable that the people also knew the secret of salt management in desert soils. This knowledge, apparently, helped them to survive for about 1000 years in this location, as archeological excavations show. During part of this time, they accomplished one of the wonders of prehistoric civilization. They built a tower 10 meters in diameter and 8.5 meters high with a staircase of 22 steps [8].

What happened in history should, however, not be derived only from the examination of remains of colossal buildings but more from what was left of the impression of ideas on man's way of life. In the case of Jericho, there are two main ideas: the domestication of plants and the development of irrigation methods. These ideas are evidenced by the carbonized grains of the domesticated wheat and barley and the irrigation ditches and furrows outside the walled town which have been reused and reploughed thousands of times since the Pre-Pottery Neolithic period. Did this revolution or, more precisely, this jump in the evolution of society, originate in one place, such as Jericho, or were there many places where man arrived at this innovation? This question, which is still debated between two schools of thought, is, in the opinion of the author, not relevant to the issues discussed in the present work. What is important is that at Jericho, where the earliest clues of domesticated grains were found, the geographical and hydrological conditions allowed only irrigated agriculture. Thus, until other evidence indicating that there was another place where such an agricultural system existed, is unearthed, the uniqueness of Jericho as a site of irrigated agriculture should be taken as an observed fact.

Who were the people who took this step? Anthropological studies on the skulls which were found show that the people were of the same stock as those who lived in this region during the Epipaleolithic period, using Natufian type flints, namely, the race or stock called Eurafricans [9]. Little is known about their spiritual beliefs. They buried their dead below the floors of their houses; they plastered the skulls which they kept in their houses, most probably as an object of cult. They had an organization which enabled them to mobilize enough manpower to build the tower, which served either as a watch tower or as a cult site, a precursor to the ziggurats of the Mesopotamian plain. They also had commercial relations with distant places, like Anatolia, from where they obtained obsidian and malachite.

After a 1000 years of occupation of the site, the Pre-Pottery culture came to an abrupt end. There are signs of flooding on top of the layers from this period. No signs of fire or other destruction could be observed. Was there a climatic change which caused the life at this site to become unbearable, or was it the pressure of enemies who made the tilling of the fields impossible? This is impossible to say. Kathleen Kenyon [8] maintains that it was indeed a period of floods. As Jericho is situated below the outlet of Wadi el-Kelt which drains a vast area of the eastern hills of Judea, a more humid period would have caused flooding and even a rise in the water table of Ein-el-Sultan, which might have caused the impregnation of the structures and even flooding of the buildings. This hypothesis still needs investigation, as only a rather small section of the layers from this period have been investigated.

The people who settled in Jericho at about 9000 B.P. had different cultural traditions from the former PPNA (Pre-Pottery Neolithic A) people. The houses they built were better constructed and the type of brick was totally different. The flint tools they used were also different and did not evolve from the PPNA as the latter had evolved from the Natufian. Today, excavations in other sites in the Levant show that these newcomers to Jericho had also previously dwelt in many other sites. They may have reinhabited the site after a few centuries when the living conditions again became bearable.

In the layers of PPNB (Pre-Pottery Neolithic B) as this new layer is termed, a building which contained a standing stone in a niche resembling a Mazaba was found. This might be a cult object, a clue to the cult practiced by the people of the place. The same can be said about the painted skulls and the three life size figures found in these layers, which comprise over 20 building levels, covering a time span of about 1000 years. The people continued to live on agriculture. The protein composition of their food was enriched and it contained an additional item of domesticated goat meat, whereas during PPNA wild gazelle was the main animal eaten. As a matter of fact, during PPNB domesticated goat meat was the main item. This finding is important as it demonstrates another economic tool which man acquired and which enabled him to penetrate into the desert areas which have spread since the end of the Last Glacial period. This is because the goat is much more adapted to desert conditions than cattle or sheep. It is thus no wonder that, while the people of Jericho ate mainly goat meat which they most probably herded in the Judean Desert west of their town or along the Arava and the foothills of Trans-Jordan, the people of Munhata in the northern part of the Jordan Valley, which is more humid, still preferred gazelle meat and wild pig from the swamp of the Huleh[10], while goat meat formed only a small part of their diet. The domestication of the goat seems also to have held another blessing. It enabled man to cope with a phase of desiccation which seems to have affected this part of the world .

The PPNB period ends abruptly at about 8000 B.P. What was the reason this time for desertion? One cannot say for sure. It does not seem that it was caused by destruction in war, although pressure by hostile nomadic people roving the country, but unable to break into the city, may have been one reason. Again, an abrupt flood or a few decades of heavy rains may have caused the people to be unable to sustain themselves. The erosion by water flow observed on the buildings of PPNB Jericho may advocate such a process. Whatever the reason, the descendants of the inventors of agriculture had to abandon their oasis and look for another site to continue their culture. It is probable that indeed a more humid period began at this time, causing the water table below the town to rise and floods from the nearby wadi to flood the fields and destroy the houses.

Thus, although the reduction in the amount of tree pollen in core samples taken from Lake Huleh in northern Israel brought the archeologist Jean Perrot to the conclusion that the climate became drier [10], it may be claimed, as a matter of fact, that this reduction is indicative of a more humid period, as man had started to cut down the forest and plough the fields for grain. Whatever the reason for the desertion of PPNB in Jericho was, it is clear that the agricultural systems which began to be practiced in the Jordan Valley, whether exclusively or simultaneously with other sites, predated the agricultural systems in the valleys on the big rivers to the north and the south. Several more centuries were needed before these systems could be expanded to the gigantic scale exercised in the kingdoms of these valleys.

Technically, there must have been a transition phase from the irrigation system with a river as a source. The people of Jericho most probably did not use the water of the Jordan, as this river flows through a deep gorge which cuts into the soft marls that Lake Lisan left behind by the prehistoric precursor of the Dead Sea. In case the water from their spring did not suffice, they could have used the springs of the Ein Duyuk and Nuweima, north of their town, or the winter flows of Wadi el-Kelt and its springs. The technical problems involved in diverting water of a rather small spring into fields which lie below its outlet are not too complicated, as large structures and long canals are not needed. On the other hand, when water is taken from a river without a pump being used, a diversion canal with damming structures needs to be constructed upstream and the gradient of the canal has to be controlled in such a way as to avoid destruction of the canal by the flow of water or loss of too much water by infiltration and evaporation due to a very low velocity of the flow in the canal.

A transition stage from spring water irrigation to river water irrigation in a semi-arid zone is found at Catal Huyuk in the Konya plain on the Anatolian plateau (Fig. 2.5). Settlement of this site began about 8200 B.P., more or less during the period of the late PPNB of Jericho.

The variety of cereals grown by irrigation with river water at Catal Huyuk exceeds that of Jericho. Wheat is the main component. The domestic animal was not the goat but cattle, which supplied the people of Catal Huyuk with more than 90 percent of their meat [10]. The variety of crops as well as the mixed anthropological types which were found at Catal Huyuk speak for the source of seeds, as well as people, from other regions. As was mentioned earlier, the obsidian found in Jericho shows that trading over wide distances was already in practice. It is thus quite possible that the people of Catal Huyuk obtained some of the seeds and know how from the people of the Jordan Valley.

A striking evolution at Catal Huyuk had also occurred on the spiritual level. Shrines where a relief of a goddess giving birth to a male lamb and bulls' horn cores set in benches and walls in all probability show that the people of this place worshipped the animal which gave them their meat and milk and most probably also their clothing and work power. Clay and marble figurines of females which were found in the houses were probably used as fertility talismans. Other female statuettes accompanied by broken stalagmites of phallic resemblance can be described as the first pages of a long story on the Goddess of Fertility or the Mother Goddess which was to be transferred from one generation to another.

At about 7300 years B.P., Catal Huyuk was abandoned and its people moved to another site in the valley. Similar agriculture based on irrigation from a river, the lower levels of which are contemporaneous with the last levels of Catal Huyuk, and which continued till about 6800 B.P., have been found in Halicar on the Anatolian plateau. The abundance of female clay figurines, some of them resembling those of Catal Huyuk, show that these people took over the gospel of the Mother who bears life.

The knowledge of irrigated agriculture at about 7600 B.P. (at the end of the Catal Huyuk early phase and contemporaneous with Halicar) enabled the establishment of irrigated agriculture in the lower and arid Mesopotamian plain. The fields in the new site were watered apparently from the Tigris and wheat, barley and flax were grown. The inhabitants kept domesticated goats and sheep and also had domesticated dogs. Female figurines made of baked clay as well as alabaster show that a fertility cult was also practiced by the first irrigation farmers to enter the fertile plain of the two rivers. Was the Mother or Fertility Goddess called Tiamat? Most probably we will never know, as these people knew only how to express their beliefs verbally and figuratively but not yet literally.

During the time of this early culture (called the Samarra culture) settlements in other places in the northern, less arid part of Mesopotamia thrived. The finds in these places do not indicate whether the people using the ceramic wares (known also as the Hassuna or Halaf cultures) also practiced irrigated agriculture. On the contrary, from the information gathered from some places, it can be concluded that they relied mainly on their herds and the natural vegetative products of the surrounding countryside. Yet the transition to agricultural settlements was approaching. But before we concentrate our attention on the settlements and cultures of Mesopotamia, we have still to read a few paragraphs on the prehistory of the very early settlements in the western Fertile Crescent.

As has been described, the cradle of agricultural civilization most probably stood in the arid valley of the Jordan, outside the walls of the Neolithic town of Jericho. This involved one of the biggest changes in the function of the system called human society, as it became a more deterministic system. How man fared became more a result of his own decisions rather than of outside factors such as migration of herds of wild cattle and abundance of wild plants. True, nature still remained the prime decision-maker. Man acknowledged it and, as an act of submission and fear, was careful to sacrifice to it a share of the fruits and grains of his fields and a few of his domesticated animals. In early times he even offered his own first-born child. Early opportunists and their followers, most probably, soon took advantage of these fears, and demanded unquestioning obedience from the frightened people, to build the early clerical and governmental bureaucracies in order to extract benefits from the working people with a minimum personal effort of their own. But let us not indulge in this story, as this is already another tale of another subsystem with its own rules.

At first, the agricultural societies concentrated in the valleys where they had a perennial source of water to irrigate their lands. Beyond the fertile valleys extended the low latitude semi-arid lands and deserts, extensive stretches of land with scant rainfall and hence sparse vegetation and life. As already mentioned, many of the valleys and flood plains themselves received rather poor rains, so the secret of survival of the settler in the valleys was man's ability to harness the water which emerged from the huge springs fed by rains falling on high, distant mountains. At an early stage man learnt to divert this water from its natural channel and irrigate the land he planted. He did this after learning from experience the basic laws governing the flow of water and thus the basic rules of water engineering. As experience continued to accumulate, knowledge continued to grow, and as the organization of human societies encompassed greater numbers of individuals, a combined and organized effort was made to push the irrigated land further into the desolate areas. At the same time some people began to learn how to manage herds of domesticated goats, and later sheep on the arid land, and in this way they may have helped man's society to benefit from the grasses and shrubs which covered the neighboring plains.

On the border of the irrigated valleys two other systems developed. One was the settled agricultural society dependent on rainwater or winter floods. This society can be described as a system that became accustomed to periodic rather than harmonic fluctuations of supply. The second was the nomadic society which had to become accustomed to random fluctuations in the abundance of sources of supply. These people learned to understand the secrets of resource management of the desert by constantly shifting their place of dwelling. It is obvious, that there existed also in-between societies which practiced a mixed way of life, namely were semi-nomadic. Contemporary observations show that as long as the natural resources were abundant the various societies lived in peace and even in some kind of symbiosis.

The periodicity built into the genetic code of the grains, which were the main sources of food, made the society dependent on this food also dependent on the seasonal calendar. The limited quantity of water available for the lands on which these grains were grown caused the settlers to become a 'measuring' society. Man had to learn to measure time in order to know exactly when to apply the water to the fields. At the same time, while making them ready to receive the water, he had to measure the area and afterwards the quantity of water which he applied to the fields. As a result, institutional organizations and conventional metric systems of space and time had to be agreed upon in order to enable numerous consumers to use one source. As already suggested, man entered the vast flood plains of Mesopotamia from valleys fed by springs and small rivers. In the big plains quantities of water and areas of lands were many orders of magnitude larger than in the small valleys. It became necessary to measure large spaces as well as slopes and large volumes of water and afterwards the crops which had to be shared among all the people who had contributed to the big projects. This brought about the invention of scales of measurement able to cope with large areas and quantities, and the mathematics and geometries which enable man to deduce from the small to the large and vice versa.

On the other hand, in the nomadic societies emphasis was put on mobility as a device which enabled man to benefit from a special system of storage characteristic to arid lands. In other words, society had to adapt ways by which to overcome the random negative changes in climate and hence in food. This they did after learning the secret that one could store enough food even in meager years from desert areas if one was able to roam the wide plains and collect the scant vegetation which succeeded in thriving there. The larger the area the more food could be stored. The storage facilities for nomadic societies, replacing the granaries of the farmers, were the fat and meat of the livestock. The number of head in a herd also symbolized abundance, strength and survival. While the number of livestock was small, water could be stored from one dry period to another in cisterns and reservoirs, but when the number increased, the use of other water sources, such as groundwater, had to be introduced. People who dared leave the safe supply of water from the rivers or springs but still did not want to be dependent on the random system of rains in arid lands had to learn the art of digging wells to find and use groundwater (see Appendix I).

The gradual development of the methods of excavating deeper wells for tapping groundwater has enabled man to penetrate deeper into the desert and thus be able to utilize the forage which the desert could supply for his livestock. In the regions which were built of impervious rocks it may be assumed that man built cisterns to collect rain water.

The spread of settlements from the Neolithic to the Chalcolithic period in the Middle East began from the lands where springs were abundant, and moved to locations of shallow water tables where groundwater could be found and easily drawn. Only later, during the Bronze to Iron Age, did man move his settlements to areas where enough water could be collected in cisterns from rain. This trend of evolution can be seen quite clearly , along the border between the semi-arid and arid zone in the southern part of Israel where, due to special hydrogeological conditions, the groundwater table is near the surface. In such a location, one can even assume that during more humid periods, such as part of the Chalcolithic, springs emerged. But when the climate became dry, man dug the shallow wells, as the water could be readily reached.

The best example is the region of the northwestern Negev through which wadis extending from the mountains of Hebron cut into the loess and sandy layers of Pleistocene and Neogene age which overlie chalk layers of Eocene age. In this area a rather large number of settlements of the Chalcolithic age have been found. The first settlements of early Chalcolithic age, ca. 6000 B.P., were close to the river-bed in places where the water emerges as springs which flow even today. During the later part of this period, settlements moved to the southeast, where the water table is lower. According to the relics left behind, the settlers continued to enjoy a high standard of living, which means they had an abundant supply of food and water. One of the sites at which the relationship between the evolution of society and of the methods of water utilization can be observed is Beer Sheva. Here in particular the evolution of the settlements is especially related to that of the development of the groundwater resources.

The settlement in the vicinity of Beer Sheva, (Figs. 1.1, 6.2) investigated by the archeologist Jean Perrot [10], started at about 5500 B.P. Subsurface dwellings were excavated by the ancient inhabitants into the loess soil close to the ephemeral river bed. According to the opinion of the present author, this shows that the climate at that time, although a little more humid, was not altogether different from that of the present-day semi-arid climate. The reason for this assumption is that if the precipitation had been much higher, the loess would have been soaked and subsurface dwellings would not have been feasible. Observations by the author show that at present infiltration is about 100 to 200 mm into the subsurface (rains in this area average 200 mm per annum). It is assumed that at the time of the Chalcolithic the maximum precipitation was double that of the present, thus the loess did not soak a soil column more than 400 mm. Moreover, the water table in the chalks and gravel below the dwellings did not rise above the level of the river-bed. Thus the artificial caves were dry throughout the year.

The reasons which brought these people to build their houses in the subsurface are not known. It might well be that they came from a region where soft rocks were widespread and where this type of dwelling was the easiest and most natural to construct. If this assumption is correct, one can even theorize that the rocks in the place in which these people first built this type of subsurface dwelling were harder and more stable than the loess soil, as the first dwellings in their new environment had large rectangular rooms of about 3 x 7 meters. These rooms collapsed and the people learnt to build small oval-shaped rooms which were more suitable to the geomechanical properties of the soil.

The people were farmers as well as herdsman. The large number of sickles and grinding stones, and the granaries show that they knew how to make use of the wide stretches of loess soil and relatively meager rainfall of this region. The topographical and the hydrological considerations make it quite obvious that their agriculture was not irrigated. The large quantity of churns found in all these sites shows that the milk products from their herds, which mainly consisted of goats, played an important role in their diet and economy. The Chalcolithic sites of Beer Sheva were occupied for a rather short time. Anthropological research shows that the people came from the north. They buried their children under the floors of their houses probably as part of a cult. It is interesting to note that many flat colored pebbles were found on the floors. The pebbles were grouped in multiples of the number seven. This is the more interesting as the name of this place "Beer Sheva" means "well Seven" . In the Bible the reasons for the name are given to commemorate the vow of Abimelech and Abraham (Genesis 21,31) and later Isaac (Genesis 26,33), as the root for the words seven and vow is the same. The findings from the Chalcolithic period show that the biblical name was a late explanation to an ancient name, adopted, apparently, by new comers.

That the number seven had a religious meaning even before the time of the Hebrews is evidenced from other excavations in Israel as, for instance, from Naharia in northern Israel, where the archeologist Dothan found a cult center of the Mid-Canaanite period (Mid-Bronze, ca.4000 B.P.) [11], where offering vessels made of seven cups were found near the altar. That the sacred significance of the number seven is connected with the lunar cult can be learnt from the chapter in the fifth tablet of "Enuma Elish". There Marduk fixes the station for the gods "The Moon-God he caused to shine forth, the night he entrusted to him, he appointed him, a being of night, to determine the days. Every month without ceasing with the crown he covered him saying: At the beginning thou shinest upon the land, thou commandest the horns to determine six days and on the seventh day to divide the crown, On the fourteenth day thou shalt stand opposite the sun." (Chap.2, [4]). The month was thus divided into four weeks each of seven days, while the last day of the old moon and the first day of the new moon were days of festivity, together making 30 days. The Hebrews though adopting the seventh day of every week as the holy day, detached it from the lunar cycle. They continued, however, to celebrate the day of the first moon. The holidays, mainly those which have a connection with agriculture, start at full moon.

A very rich culture of the Chalcolithic period was also discovered at Tuleilat Ghassul on the eastern side of the Jordan rift valley southeast of Jericho. Today, this site is devoid of any water resource. It is thus quite obvious that the flourishing of the Chalcolithic settlements was due to a more humid climate which prevailed in the

Fertile Crescent. In order to see the extent of this change over all the Fertile Crescent, the correlation chart between the various diagrams of the isotopes can be reexamined (Fig. 1.3). The depletion in the environmental isotopes at the lowest part of the curve speaks for a colder climate during the Chalcholithic Period. Could this colder climate in the Anatolian Plateau have driven the people of Armenoid type southward, as evidenced from the anthropological investigation of the skulls found in Beer Sheva? This should be taken as a hypothesis. It is quite clear, however, that the climate during the Chalcolithic period was more cold and thus humid, and man could penetrate deeper into the desert with his livestock.

This humid climate also enabled people to live in the Judean Desert west of the Dead Sea (Fig. 4.1). This area is arid due to its location in the shadow of the rainstorms coming from the Mediterranean Sea in the west. Today the bedouins graze their goats, sheep, and camels in this area only during the winter and spring when grasses are abundant. It is very probable that at the humid periods the vegetation was more abundant. There exists, however, the problem of a perennial supply of water in this area due to the geological and topographic conditions which do not favor the formation of groundwater aquifers at shallow depths. Thus it seems possible that the water was supplied in the same manner as today, namely collection pits, cisterns, or caves excavated into the soft chalk in the vicinity of the river-beds. This, however, could not supply permanent settlements. These were located in the vicinity of springs or the shallow water table along the peripheries of the desert, either to the west along the watershed or to the east along the shores of the Jordan and Dead Sea. One of these sites is near the spring of Ein Gedi (Fig. 4.1), where a cult center has been excavated. This center is composed of a big yard and a building in which an altar on which many burnt bones of livestock were found. A clay statue of a bull carrying two butter churns was also found in this temple [12]. Many remains of the inhabitants of this period were also found in the caves in the canyons leading to the Dead Sea. Of special interest is a treasure composed of bronze implements and cult objects that were found in one of these caves. It seems that they were hidden there to protect them from an enemy or when the people had to move from this area [13].

5 The Two Rivers, Givers of Life and Law

> And it came to pass, as they journeyed from the east, that they found a plain in the land of Shinar; and they dwelt there. And they said to one another, go to, let us make bricks, and burn them thoroughly; and they had brick for stone, and slime had they for mortar. (Genesis 11:2,3).

As has already been mentioned, the lower stretches of the Mesopotamian plains are arid, and thus only irrigated agriculture can be practiced there. Thus the settlement of this region by land-tillers started only after man learnt to build canals and divert water from the big rivers. This happened during the Late Neolithic-Early Chalcolithic, (namely, the Hassuna-Sammra periods at ca. 7500 B.P.). During the Late Chalcolithic one finds a profusion of settlements, all representing a well-defined culture (called Ubaid) which began at about 6900 B.P. and continued till about 5300 B.P. The people who practiced this culture most probably came from the southeast where they developed methods of irrigation from streams and springs flowing in the valleys of the Iranian plateau and later at about the middle of the Chalcolithic period, they arrived at a stage which enabled them to cope with the technical and organizational problems of irrigation from the big rivers on a large scale. As stone is rare in the Mesopotamian plain, these people made full use of the clays. Their pottery and bricks and also their sickles, axes, and knives were of burnt clay. Their Mother Goddess was also made of clay. The temples were built one on top of the other.

The first people who began to develop their civilization (called The Ubaid culture) in the land which was to be called Sumer (Biblical Shinar?) may have belonged to the aborigines of the valleys of Iran before the penetration of the Indo-Aryans. They settled in the flood plains of Mesopotamia and began to practice irrigation from the rivers to supply their rather large communities. They might have learnt these methods also from their neighbors in the northern part of Mesopotamia or may have brought them from the valleys extending from the Iranian Plateau. The beliefs and cults they brought with them from the east came in contact with those of the Semites from the west and of the aborigines, who may have occupied the northern part of Mesopotamia, since Neolithic times, as shown by some ancient Mesopotamian place names which are neither Semitic nor Sumerian.

The Ubaid culture was replaced by the Uruk culture, named after the city of Uruk (the Biblical Erech) where it was first unearthed. At Eridu, another Sumerian town and one of the most sacred, as it was the residence of Enki, the God of Waters, 17 successive layers containing ruins of temples were found. Plans of the temples are identical. In each one of the layers of ruins a thick layer of fish bones is found on the floor of the temple, apparently food brought to Enki. This succession of layers continues without interruption into the historical period.

During the fourth millennia B.P. (during the Uruk period) came the invention of writing. It developed into the cuneiform script which served the literate world until the end of the Achaemenide period, namely the time of Alexander the Great of Macedonia (ca.300 B.C., 2300 B.P.).

The civilization which started in the southern part of Mesopotamia was organized in large communities; in the beginning in city-states, later in kingdoms which dominated vast areas, and finally expanding into empires. Although, as can be seen from the archeological record, the colonization of northern Mesopotamia had already begun in the Lower Neolithic period and reached southern Mesopotamia only during the Upper Neolithic-Early Chalcolithic, in later periods the people of the south ruled the north. This was most probably due to the fact that the people of the south had to organize bigger governmental systems in order to cope with the basic problems of controlling their irrigation systems, while the people of the north could make do with the natural supply by precipitation. The irrigation systems brought with them the need for an administration supported by a militaristic system which had to enforce taxes, build, control and protect the water projects.

Thus the Sumerians, who came most probably from the east, occupying the southern part of Mesopotamia, and the Semites,who came from the west, and north-west, occupying the central part, shared the same big valley and were in close contact. They exchanged world views and traditions, while developing their written literature. It seems quite obvious that they started from different archaic, most probably prehistoric, religions and cultures, as already told, and as will be discussed later in more detail. But living together in the same valley and sharing the same environmental conditions caused them to develop a similar culture. As was demonstrated in the chapters dealing with the myths of creation and the Deluge, the Akkadians adopted quite a lot of the earlier Sumerian myths, but they also brought some of their own with them, which they changed to fit better into the framework of the new environment. It seems that the analogy found between the myths in the Akkadian and Biblical texts is due, in the first place, to a common pre-Mesopotamian heritage, but also to a blending process which took place especially after Sumer was conquered by the Semites, as will be told later on.

The Sumerians were the first who managed to build large water projects connected with their ability to build the first big organized settlements. In about 5000 B.P. Uruk had a population of about 50,000 inhabitants while Old Uruk in about 4800 B.P. had about 34,000 [1]. The pivot of the organization was the local god, its temple, and the religious ceremonies which took place according to the changing seasons. The king and ruler of the city was a vassal of the divine power.

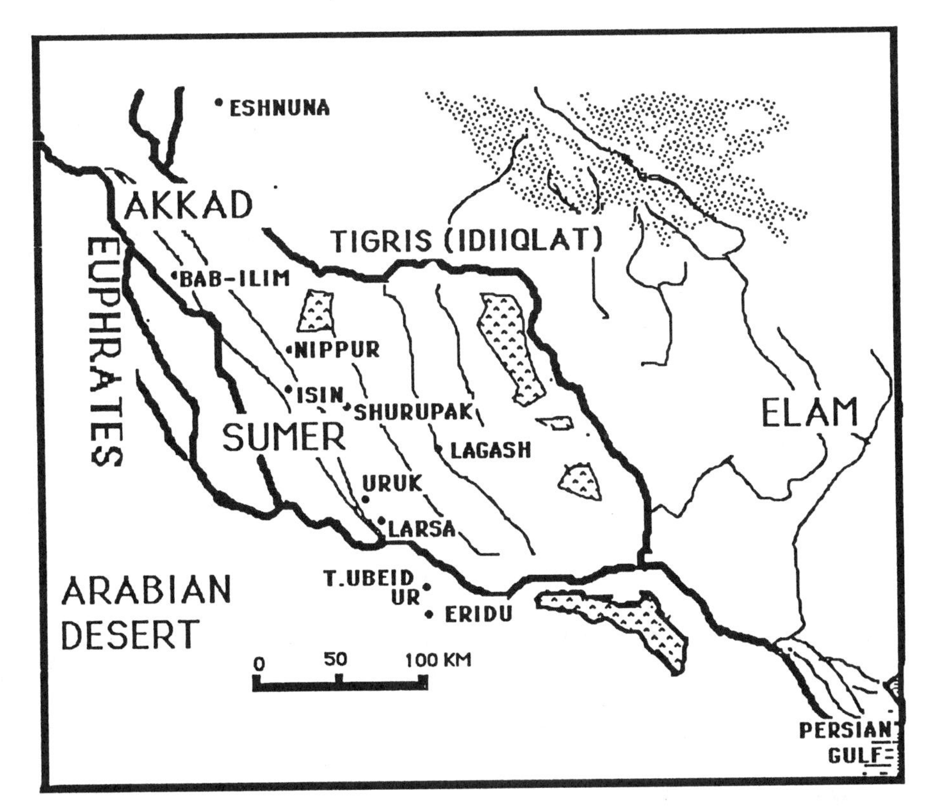

Fig. 5.1. Map of Sumer and Akkad

One of the most important aspects of the ancient world which developed in Sumer in connection with the big irrigation projects was slave labor. The larger the public projects became, the more man-power was needed for their execution and maintenance. The hierarchical pyramid became more and more established as base layers became more numerous, its top richer and the in-between layers more dependent on the richness of those at the top and the work and subordination of those at the bottom. This brought about the evolution of hierarchical and despotic governmental systems characteristic of the big river civilizations.

A rather short cultural transition time (named the Protoliterate period) marked the last steps of emergence from prehistory to history. As the invention of writing spread, the Uruk culture was replaced by the Sumerian Early Dynastic kingdoms, the history of which was written on the cuneiform tablets found in the excavation in the Mesopotamian Valley.

As already mentioned, while discussing the date of the Deluge, it may well be that during the Protoliterate or even the late Uruk times (around 5000 B.P.) a climatic change took place which produced a series of floods which destroyed towns and fields and was afterwards explained and narrated as a global catastrophe.

The historical records found in Sumer start the list of kings from the Deluge on. Among these kings was also Gilgamesh, the King of Uruk, whose search for immortality brought him to hear the story of the Sumerian Ut-Napishtim, the survivor of the Deluge. Thus, at about 4750 B.P. the period of the Sumerian city-states ruled by the post-Deluge kings or, as it is also termed, the Early Dynastic Period started.

The Land of Sumer was an agglomerate of city-kingdoms, strung along the Euphrates from the latitude of present-day Bagdad to the Persian Gulf (Figs. 1.2, 5.1). Each city-kingdom was surrounded by satellite towns and villages situated between fields ("gan" in Sumerian), orchards and palm groves irrigated by canals from the river. The open uncultivated steppe between the irrigated areas served as pasture land and was called in Sumerian "edin", thus natural noncultivated stretches of land were called "Gan-Edin" in Sumerian. This reminds one of the biblical word "Gan-Eden" (Garden of Eden) for the archaic garden where man lived so happily before he had to toil on his land for his bread. The pivot of each city-kingdom was its god-protector. Thus, Enlil, the chief Sumerian god, was the patron of Nippur. Uruk was dedicated to Ann (Anu), the sky god, and Inanna (Semitic Ishtar), the Goddess of Love. Ur belonged to Nanna (Sin of the Semites), the Moon God; and Lagash to Ninqirsu, son of Enlil. The god of Larsa was the sun god, Utu (Semitic Shamash). Enki, whose domain was complex, was the protector of Eridu. Literally, his name means "lord earth", but possibly because the earth without water had no use, Enki also became the deity of the sweet waters. At the same time he controlled the flow in rivers and canals, the rise in springs and wells. The cultivated land around the cities were divided into three parts: one part, the land of the lord (gana-ni-en-na), was worked by all the community for the sole benefit of the temple; another part, the food land (gana-ku-ra), was allotted to the dependents of the temple; the third part, the ploughed land (gana-apin-la), was let to tenants against one seventh or one eighth of the harvest. Thus, the temples had at their disposal high revenues which were used also to finance large-scale operations, building of temples, fortifications, and irrigation canals. The ruler of the city-kingdom was thought to be chosen by the city-god and was, in the archaic period, also the chief priest. He was also considered to be the personification of the god. Afterwards the tasks were divided between king and priest, but the king (Ensi or Lu-gal) was the one responsible before the god for the well-being of the people of the city (mainly Chap. 3, [3]).

War between the cities was for possession of the irrigated lands. Success and victory meant many irrigated lands, revenues and power. Thus after a period characterized by the rule of city kings, victorious kings ruled over many cities. The Ensi of Umma, at the command of his god, Enlil, raided the irrigated land, the fields beloved of Nir-gir-su. He ripped out the stele which marked the border between fields, cities, and kingdoms, and entered the plain of Lagash. The King of Umma went on to build a state kingdom, he conquered all Sumer and claimed even to reach beyond its border: "When Enlil, king of all sovereign countries had given him the kingship over the nation.... Enlil let him have no opponent, all sovereign countries lay (as cows) in pasture under him; the nation was watering its field in joy under him" (Chap. 3, [3]).

This Sumerian kingdom did not last for a long time. Sargon the Akkadian King of Kish conquered Sumer in 2371 B.C. and established an empire from Iran to the Mediterranean, ruled by the Semitic Akkadians. This empire, which lasted, although in a reduced area, until 2230 B.C., caused the expansion of the Semitic language and cuneiform script over most of the Middle East. Trade and exchange of merchandise, as well as technologies, extended from the Hindus Valley and plain (where another wonderful irrigation civilization emerged), to Anatolia, Greece, and Egypt.

The second Sumerian period started about 2100 B.C. after an interval of about 100 years of domination of the Mesopotamian valley by outside conquerors, the Gutians. During this Neo-Sumerian reemergence, the city of Ur under King Ur-Nammu seized hegemony over the other cities. This empire later expanded beyond the borders of the valley, governing a vast territory by local En-sis, each Sumerian or Akkadian ruling as a vassal to the great king. This period, which was peaceful inside the empire, although war persisted along its northeastern borders, was the period of the building of the gigantic ziggurats, an echo of which one finds in the biblical story of the Tower of Babel.

In the year 2006 B.C., Ur, the capital of the Sumerian kingdom, was destroyed by the Elamite tribes living along the Zagros foothills. This destruction was the end of Sumer as an authentic independent society and culture. The stage was cleared for another society of western Semites. These were called by the Sumerian Mar-to or Mura-tu (Amorites), namely "westerners". They came out from the more desert part of the Fertile Crescent and penetrated into Mesopotamia, conquered many cities and settled down in and around them. They adopted the parts of the Sumerian beliefs and ways of life which were to their liking and which they could make use of. Although the coup de grace to the capital Ur was given by the Elamites, it was only a long process of deterioration of the kingdom and the continuous encroachment of the Mura-tu which finally brought about its collapse. The history of Sumer became buried under the ruins of its cities, turning into clays due to the disintegration of the mud bricks.

The downfall of the Sumerian kingdom as a result of the invasion of the people from the west is well documented by clay tablets found in the archives of Mari and Ur. They portray the sequence of events which has recurred many times in history: the collapse of a well-established, sedentary, civilized society under the attack of "noncivilized" nomadic and semi-nomadic tribes. The latter being dependent on the random supply of rains, once this supply fails, their only possibility to survive is to encroach on the fertile plains irrigated by the water of the rather stable big rivers, where, food, pasture, and other trophies were abundant. As already mentioned, nomadism, usually interconnected with the sedentary communities near springs, started to evolve as a way of life after the goat and sheep became domesticated, probably during the Neolithic period . The need to cover large areas in order to supply food for the livestock brought about the formation of a new type of society which existed side by side with that of the societies living on irrigated agriculture. The relations between the two depended on the climate. When this was favorable these two societies complemented each other from the economic point of view (and sometimes even the social by providing mercenaries); but once a period of droughts encroached over the desert, the nomadic society had no other resource but the fertile fields of the sown lands. When they could not obtain this land against cash or kind, they took it by force.

This clash between the pastoral nomad and the farmer is a story which repeats itself throughout the history of the Fertile Crescent. Its first echo is heard in the biblical story of the conflict between Cain and Abel. It might have started with the drying up of the climate in the later part of the Neolithic, around 6500 B.P. This caused most probably the penetration of agricultural societies into the more southern parts of the fertile plains of Mesopotamia which were practically empty. An indication of another aridization phase also at about 5000 B.P. (3000 B.C., Mid - Early Bronze Age) and another one at about 4000 B.P. (2000 B.C., End-Early to Mid. Bronze Age) can be detected in the isotope and pollen curves discussed in the preceding chapters (Fig. 1.3). In these curves, one can see that, at about 4000 B.P., as the ^{18}O and ^{13}C become more abundant, there is a simultaneous reduction (in the core samples taken from the bottom of the Sea of Galilee) in the pollen of olives and increase in that of oaks. These changes apparently show that, as the semi-arid environment of the Sea of Galilee became drier, man abandoned the olive orchards and the natural forest re-established itself. This regional aridization might have caused the nomadic and semi-nomadic societies to try and find refuge in the green and fertile valleys. This time, however, other than in their earlier encroachments, they met well-established societies, the people of which were not willing to share their food and possessions with the newcomers. Thus a series of clashes and wars started, which caused the downfall of the Sumerian kingdom and the seizure of power by the invaders. In most cases the invaders became the ruling class, divided the lands, temples, and treasures of the deposed former ruling class among themselves and slowly assimilated with the sedentary population.

These conquering people came from the west and, as Semites, had a heritage different from the Sumerians but similar in many aspects to that of the Akkadians. This Semitic culture evolved outside Mesopotamia during the Chalcolithic to the end of the Early Bronze Age (from about 5500 B.P., 3500 B.C to about 4000 B.P., 2000 B.C). We will return to these people later; for the time being, in order not to interrupt the story of the societies of the Mesopotamian plain we will continue with the Amorites. As already related, they succeeded in establishing themselves in southern Mesopotamia, and seized power and control over the sedentary societies. Thus a new ruling class of kings and En-sis of non-Sumerian origin established itself in Sumer. These people soon adopted the Sumerian-Akkadian culture. As their language was similar to the Akkadian, they went through a process of assimilation which led to the formation of a Semitic-Sumerian culture which succeeded in replacing the original Sumerian, although it adopted its script and many of the customs.

Sumerian deities and myths were adopted and blended into Semitic mythology and rites. The tasks and dominion of the gods were changed here and there to agree better with the original Semitic gods, whose names were maintained. These gods, after whose names the rulers were called, presented another agricultural system dependent more on rain than on flood, thus the name of the god Adad, who was the controller of the rain, became prominent beside Ishtar, Dagan, and El.

A series of waves of repeated invasions continued, following each other. The newcomers, driven out of their country by the continuing famine, tried to gain their share in the spoils of the rich lands. However, after replacing the ruling class and becoming established economically, the first wave found more interests in common with the original society over which it was ruling than with the new arrivals . After two or three generations the Amorite, Sumerian and Akkadian found themselves in need of fighting back the new waves of invasions by people from the west. Thus, between about 4000 and 3900 B.P. (2000-1900 B.C.), Mesopotamia went through a series of invasions and establishment of small kingdoms ruled by the descendants of the newcomers who fought back the new invasions of Amorites and afterwards fought each other. Each tried to encroach on the lands of the other, until one of the new dynasties succeeded in overpowering the others and establishing a new empire. The man who brought this process to its climax at about 3800 B.P. (1780 B.C.) was Hammurabi. Before we turn to this man and his Babylonian empire, another small kingdom, namely that of Larsa, and especially its archive of public waterworks, has to be mentioned.

The kingdom of Larsa was established by an Amorite chieftain by the name of Nablanum in the year 2025 B.C. After four generations his descendant, Gungunum, in the year 1924 B.C., overpowered Lipit Ishtar, the ruler of the adjacent kingdom of Isin, which was established by another Amorite chieftain. The annexation of Isin enabled the establishment of a new empire which extended over the area which comprised most of ancient Sumer. During the reign of the son and grandson of Gungunum, the kings Abisare (1905-1895 B.C.) and Sumuel (1894-1860 B.C.), the building of new irrigation projects and the maintenance of others was carried out. A description of these works is found in an archive of official records which consists of letters and documents written on clay tablets in the Akkadian language with many Sumerian terms [2]. The tablets mention mainly a certain official by the name of Lu-igisa who is entrusted with certain public water projects and has to administer the works, payment, protection, and even surveying of new canals. The name of this official is Sumerian (meaning "a man with a kind face" or "the friendly benevolent one") but many of the other names, especially his superiors, (not to speak of the king, Sumuel) have Semitic names such as Nur-Samas (the sun gods of the Western Semite) or a lady called Arnabum ("Bunny" or Arnevet in Hebrew). From the translated tablets [2] it seems that although the governing class was Semitic, the administration, irrigation and farming traditions remained Sumerian.

At Larsa, the Irrigation Bureau was headed by Nur-Sin, who was responsible to King Sumuel in all matters regarding the supply of water for irrigation. Under him served Lu-igisa, who seemed to have been a man for all purposes. A few of these records are presented here for the reader to share the echoes of ancient times.

"To Lu-igisa speak. This is what Nur-Sin says: The side canal which I and you surveyed, the son of Ku-Nanna is about to excavate. Before he begins, you excavate it and report to Isar-kubi." ([2] p. 33).

"To Lu-igisa speak. This is what Nur-Sin says: Isar-kubi has just written me about emergency work needed at the Nubitar Canal. Here is his request: Hire 1800 workers to take care of you. Ask around, and pay ten "mana" of silver in order to hire workers. And until you can get away appoint some one of your choosing to supervise the crew." ([2] p. 35).

Other people with familiar names are mentioned in other tablets for instance:

"To Banum speak. This is what Abaraam says: You need to know that E-Kibi and Nabi-Enlil have just sent 5 shekels of gold intended for the canal to Isar-kubi in order to eat your grits (take advantage of you). Take action or Isar-Kubi will confront you." ([2] p. 109).

From the other tablets a rather clear picture emerges of how dependent the people of the kingdom of Larsa and, as a matter of fact, all the peoples of southern arid Mesopotamia were on the irrigation network. These canals were the main arteries which supplied the water for irrigation. They had to be dug and continuously cleaned so that they would not become blocked by silt brought by the rivers. This was the task of the administration. At the top of the hierarchical ladder was the king, Sumuel. It was he who decided about the big project as well as allocation of the funds needed for it. He was also the arbiter and judge to whom the heads of projects had to report or submit their claims: "If you think I have responded falsely I will even respond to Sumuel my lord." ([2] p. 107). The chain of hierarchy under the king started from the head of the Irrigation Bureau, Nur-Sin, who was most probably an Amorite, inspectors who had Amorite as well as Sumerian names like Lu-igisa, labor contractors, suppliers, foremen, workers and soldiers. The work of the Bureau for Waterworks consisted not only of digging and cleaning the canals but also preparing some fields for irrigation by furrowing them, and seeing that money allocated for work contracted should not be misappropriated.

Thus the archive of Larsa, as well as other documents such as the Laws of Hammurabi, which will be discussed later, portrays all the features of a sedentary, established society based on river irrigation practices. This only about 100 years after the invaders took control. It can be concluded that the Amorite people, although a minority, were clever enough not to destroy the old system but to get the best out of it by adopting its administrative organization and methods, though changing its language as well as adding new ideas, gods, and traditions to the existing cultural world. After they settled down they, too, became a river plain society.

It seems, however, that not all the people of western Semitic origin settled down. From the stories of the Bible we can gather that Semitic people with names similar to those we find in tablets of Larsa travelled in the opposite direction, namely, from

Ur in the southern part of Mesopotamia to Haran on the upper Euphrates (Fig. 5.1) and from there to Canaan and southward into Egypt. What were the reasons for these wanderings? Was it a second wave of Western Semites who found the country already taken and occupied by the people of the first wave and the gates of the fertile land closed? Was there a possibility that the people who settled down did not like the ways of a sedentary life and longed for a life of travel without bondage to the king and his administrators? Or was it an economic reason, namely that the country conquered was not able to support the newcomers in addition to the existing population, or may it have beeen an environmental crisis, such as drought or soil salinization, which affected the population of the Mesopotamian valley? The explanation suggested in this work is that the main reason was aridization. This, however, will be explained later on. Staying in Mesopotamia for the time being, it is worthwhile discussing, in more detail, the prevailing theory, mentioned already in Chapter 1, that the gradual salinization of the soil due to over-irrigation led to a decline in production and thus to an economic catastrophe. This explanation, suggested by the archeologist Jacobsen [3, 4], is mainly based on three main sources of evidence:

1. Records from clay tablets from this period on salinization.
2. The shift from wheat to barley. Whereas in 2500 B.C. wheat was grown in equal amounts to barley and as late as 2400 B.C. it was still being raised in the proportion of 1:6 to barley, wheat accounted for less than 2% of the grain crop in 2100 B.C. and is missing altogether after 2000 B.C. As barley is more salt- tolerant, Jacobsen argues that salt encroachment caused this shift.
3. Reduction in crop yields. Estimates based on the records found in the Sumerian and Akkadian tablets, show that in 2400 B.C. the yield was 2537 liters per hectare, in 2100 B.C. the yield was 1460 liters per hectare and in 1700 B.C. only 897 liters per hectare.

Accordingly, Jacobsen suggests that the change from wheat to barley and the reduction in the yields was the result of salinization. This happened, in his opinion, due to the addition of large quantities of irrigation water to the area,which caused the water table to rise and come near the land surface, then by the sucking upward through the fine pores of the clayey soil reached the surface, evaporated, and left its salt behind.

Although, in principal, this anthropgenic process may have happened, it is still questionable whether it caused the reduction in yield or whether there were not some other natural processes that caused the agricultural system of Mesopotamia to decline. The biggest decline in yield was from 2100 B.C. to 1700 B.C. which, as mentioned already, was also the time of a dry period which affected the entire Middle East and may have caused a fall in the total yield of the rivers and thus in the availability of water for irrigation. Moreover salinization due to a regional rise in the water table is connected with quite clear environmental phenomena on a regional scale, yet there is no mention of these phenomena in the Larsa water archives. If this process had been as crucial as suggested, one would expect to find it being mentioned more frequently and specifically in these archives which, as was shown, specify problems and methods in much detail. The alternative reason suggested by the present author, namely a severe reduction in the flow of the rivers due to desiccation of the climate, which brought a decrease in the amounts of water available for irrigation in addition to an increase in demand as a result of increase in population, may have caused the same environmental results. These will be local patches of salinized soil, as well as a shift from wheat to barley, which is drought-resistant as well as salt-resistant. The reduction in yields may also be due to smaller amounts of water applied in irrigation. It is thus suggested that, until more evidence is found in ancient documents, the blame laid on the Mesopotamian people for their misuse of the water and the spoiling of their land should be taken only as a hypothesis rather than as a proven fact. (See also note at the end of this chapter).

The names of the kings and people of Larsa are quite familiar to those who know the Bible. An even more familiar connotation can be found in the Laws of Hammurabi: as mentioned above, the waves of Semitic-Amorite tribes encroaching over Mesopotamia continued also after the people of the first wave succeeded in establishing themselves as ruling dynasties over the people of Sumer. In the first year of Sumuel, King of Larsa (1894 B.C.), another Amorite chieftain by the name of Sumu-Abum established a small kingdom in the Sumerian town of Ka-Dingir-Ra which was translated into Akkadian as Bab-ilim, namely, the gate of gods, or Bab-El, the gate of god. (The reader is reminded that in the Bible the reason behind this name is different and is connected with the language of the builders of the Tower of Babylone, being confounded, Genesis 11,9). The descendants of Sumu-Abum, slowly widened the domain of their country until in 1793 B.C. Hammurabi (Hammu - a Semitic deity, Rabbi - great) ascended the throne. In a campaign which lasted 38 years, he was successful in annexing all the kingdoms from Assur and Mari in the north to Larsa in the south.

The time of Hammurabi was also the time in which the Babylonian or Akkadian civilization was established. This was done by amalgamating the different religious traditions and laws into a state religion and civil code. Marduk was proclaimed the chief god and was accepted by the former gods as their savior from the forces of chaos and destruction. The story of creation, the Enuma Elish, already mentioned, as well as the other myths and traditions, were adopted and codified.

The symbols of order were, of course, the laws, religious as well as civil. The famous code of Hammurabi contains many laws similar to those which are found in the Bible. The detailed comparison between the two systems of laws is beyond the scope of the present study, which concentrates on the hydrological aspects of history, yet it will be shown later, that though the Hebrew law was in some ways influenced by the Babylonian one, the two systems represent two altogether different socio-economical environments. At the same time, one gains from the Laws of Hammurabi interesting glimpses of the life of the people in Mesopotamia which are important in order to understand the civilization which had still much in common with that of the Hebrews.

The interesting aspect from the hydraulic point of view is that while the laws dealing with criminal behavior of the citizens have parallels in the code of Moses, those which deal with land do not. Thus one can assume that while the criminal code was part of the Amorite-Semitic heritage the land code was adopted from the ancient Sumerian law.

Moreover, the criminal laws (Chap. 3, [10]; Chap. 5, [5]) which are most similar to those found in the Bible are feasible in a nomadic society as well as a sedentary one, while the land and water laws are, of course, suited to a farming community only. Also some of the criminal laws show the influence of a river society. Thus, while the first law is the following: (1) If an awelum (citizen) accused another awelum and brought a charge of murder against him but has not proven it, his accuser shall be put to death. Such a law can be compared to that of Moses' code "But if a man come presumptuously upon his neighbor, to slay him with guile, thou shalt take him from mine altar, that he may die" (Exodus 21:14); on the other hand, the second law of Hammurabi states the following: (2) If an awelum brought a charge of sorcery against another awelum but has not proven it, the one against whom the charge of sorcery was brought, upon going to the river, shall throw himself into the river and if the river has then overpowered him, his accuser shall take over his estate, if the river shows the awelum to be innocent and he has accordingly come forth safe, the one who brought the charge of sorcery against him shall be put to death, while the one who threw himself into the river shall take over the estate of his accuser.

It is quite obvious that two different world views, stemming from two different environments, can be seen here, the first of which may be the law of any society. Its resemblance with that of the Hebrews may point to a Semitic origin. The second law is that of a society where "the river", in this case the Euphrates, is regarded as a divine entity which decides the fate of man.

Other examples of the sort of laws found in that of Hammurabi but not in the Bible are the following:

(53) If an awelum was too lazy to make the dike of his field strong and did not make his dike strong and a break has opened up in his dike and he has accordingly let the water ravage the farmland the awelum in whose dike the break was opened shall make good the grain that he allowed to be destroyed.

(54) If he is not able to make good the grain, they shall sell him and his goods, and the farmers whose grain the water carried off shall divide.

(55) If an awelum upon opening his canal for irrigation became so lazy that he has let the water ravage a field adjoining his, he shall measure out grain on the basis of those adjoining his.

(56) If an awelum opened up the water and then has let the water carry off the work done on a field adjoining his, he shall measure out ten kur of grain per eighteen tha.

These were the most obvious cases which show the Mesopotamian sources of the Laws of Hammurabi. Also the other laws which have nothing to do with irrigation and water show quite clearly that the suggestion of seeing the code of Hammurabi as a precursor of the code of Moses does not take into consideration the basic differences in environment, society, and spirit, not to speak of the religious aspects of the two codes of laws.

The personal involvement of Hammurabi in the technical problems of water supply can be learnt from a few letters written on clay tablets in which he gives detailed instructions to dredge and clean certain canals, in order to secure the income from the rented fields. In one of the letters he instructs an official by the name of Shamash-Azir (helper of the Sun God) to investigate a complaint of a tenant that the level of a certain field is above that of the canal and cannot be irrigated. In case the complaint is true, he should be given another irrigable parcel belonging to the fields of the Palace, and "let him have no reason to withhold the barley which he pays for rent" [6].

The death of Hammurabi (1750 B.C.) was also the beginning of the end of his Babylonian empire, caused by internal as well as external forces. The people of the former kingdoms of Mesopotamia revolted and on top of this there followed invasions by peoples from the east and northwest. The invaders this time were of Indo-Arian origin, namely Kassites and Hittites.

In the year 1595 B.C. Babylon was conquered by the Hittite king, Mursili I. The king, Samsu-ditana, was killed and the statue of Marduk was taken as booty. After the Hittites left Babylon, most probably because of internal palace intrigues against the campaigning king, Babylon came under the rule of the Kassites, who later promoted another Babylonian-Kassite empire to which we will return after surveying the developments which took place in the western Fertile Crescent since we left it at the end of the Chalcolithic period.

NOTE: At the very last stage of preparing this book for press, the author received a reprint from the marine-archeologist Dr. Avner Raban, in which he summarized observations on the Bronze Age settlements along the Israeli coast [7]. In addition to his archeological findings, Dr. Raban reports that around 2000 B.C. there occurred a rise in the sea level and a rise in the supply of silt and sand which caused the silting up of the river outlets. This forced the people who used the rivers' outlets as ports, to excavate artificial outlets. The rise in sea level and an increase in the supply of sand (see discussion in Chapter 1) supports the present author's argument regarding a global aridization phase around 2000 B.C.

6 Water in the Land of El and Baal

> Give ear, O ye heavens, and I will speak; and hear, O earth, the words of my mouth. My doctrine shall drop as the rain, my speech shall distill as the dew, as the small rain upon the tender herb, and as the showers upon the grass. (Deut. 32:1,2)

We left the Jordan Valley sometime at the end of the Chalcolithic period, namely at about 5000 B.P. (3000 B.C.), and concentrated our attention on the evolution of the hydraulic society of Mesopotamia. We learnt that in about 4000 B.P. (2000 B.C.), a wave of western Semitic invaders swept into the valley and imposed its reign on the local people. The explanation given in this work, for this process, is a desiccation phase which came over the Fertile Crescent and which caused the people living in its more arid areas to seek refuge in the irrigated lands.

Who were these people and what do we know about the role of water in their fate and faith? For an answer to this question it is suggested to step down a few steps in the ladder of time and see what happened in the period after that of the Chalcolithic period in the region between the Euphrates and the Mediterranean.

In the Sumerian and the Akkadian tablets these people are called Muratu or Martu which meant "westerners" [1]. They were Semites composed of a few subgroups coming from different parts of the "west" in a series of waves and establishing themselves in Mesopotamia.

There is no direct evidence for establishing the origin of the Semites whether they were endogenous to the region between the Mediterranean and Mesopotamia or immigrants from another region. Most of the existing evidence, linguistic, archeological, as well as ethnic [2], suggests that they came from the Arabian Desert, although it is quite obvious that absorption of other ethnic groups, most probably northern ones, took place during the Chalcolithic period. As was pointed out earlier, the settlements of Beer Sheva show northern anthropological characteristics. It can be assumed that the core of the Semitic civilization evolved from the prehistoric societies which had dwelt in the Fertile Crescent since the Upper Paleolithic, if not earlier, and developed a unique civilization through a process of evolution, absorption of marginal societies, as well as adaptation to invading groups.

The uniqueness of this civilization is quite obvious when its world-view is compared with that of the neighboring ones, especially in all that is concerned with the forces of nature, defined by its mythological traditions, although as one can expect, there are many similarities, due to transfusion processes, such as those which we have already witnessed and which exist between the Mesopotamian and western Semitic mythologies.

The most revealing and authentic information on the Semitic world of beliefs is derived from the tablets found in Ugarit, situated in Syria on the Mediterranean coast. These were written in about 1300 B.C., but transmit a much earlier mythology of the people of this region. In the coming paragraphs this mythology, in particular the parts concerning the influence of climate and water, will be surveyed.

El, "The God" headed the Pantheon, but his power was rather limited. The executive powers were seized by his descendants Baal and Anath. The palace of El was situated at a great distance, a thousand plains, ten thousand fields from Canaan at the source of rivers in the midst at the outlet of two deeps (Teomotaim). It seems very probable that this text revealed the most ancient source of Semitic religion in the deity of the water which emerges from caves fed by subterranean rivers. El was also called Abu Shanima meaning the father of years, or time. This name may be connected to the fact that many of the springs in the limestone terrain of the Middle East are pulsating springs, which means that there is a change in their flow every few hours as a result of pressure building up in subsurface caverns. Such a spring was Gihon of Jerusalem, from which water was taken to the temple of Jerusalem and at which the kings of Judea were enthroned. Springs in the Fertile Crescent also undergo periodical change due to climatic changes from season to season, especially from winter to summer. In this respect El has parallel characteristics with Cronos, the archaic head of the Greek Pantheon who was dethroned by Zeus, as was El by Baal, although the latter retained his authority among the gods. According to the Ugaritic tablets, El seduces two women and allows them to be driven into the desert after the birth of two children "Dawn" (Shahru) and "Sunset" (Shalmu). El is also named "The Bull God" (Tor El), although the name El or "Ilum" is nearer to the name of the "Ail" or mascot.

The main active and worshipped divinity in the Western Semitic Pantheon was the storm god Baal. The word Baal means lord, but its archaic meaning may be ba-al, namely, "came-upon". Baal was also worshipped as Hadad (Akkadian Adad) who, after the 15th century B.C., became the main god. As the chief of gods, Hadad, the lord of storm, lightning and rain, was enthroned on a lofty mountain in the far northern heavens and was sometimes considered as the "Lord of Heaven" (Baal Shamem). As the god of the storm, whose voice resounded through the heavens in the form of thunder, he was a giver of all fertility [3]. In the tablets of Ugarit, Baal is called "the Son of Dagon". The later deity is of a complex origin, the name itself has a connection with fish (Dag), but at the same time it has in Hebrew the meaning of grain. Dagon was one of the oldest Akkadian gods, worshipped over the entire Euphrates Valley as far back as the 25th century B.C. The author suggests that the connection and change from the domination of El sitting "on the source of two rivers in the midst of the fountain of two deeps" to Baal-Hadad, god of storm, lightning, and rain, who is also son of the god of grain (or other staple food such as fish), may be explained as the transition from a way of life which was based on water flowing from springs to a dependency on rains.

Another symbolic fight, which reminds one of the fight between Marduk and Tiamat, is that between Baal and Prince Yam (Sea), also called Judge Nahar (river). "The club swoops in the hand of Baal like an eagle between his fingers. It strikes the pale of Prince Yam between the eyes of Judge Nahar. Yam collapses, he falls to the ground, his joints bend, his frame breaks" (Chap.3, [10]). Prince Yam-Nahar is the beloved son of El, who ordered Kothar-Wa-Khasis (later pronounced Koshar), the wise craftsman, to build a palace for Yam-Nahar on El's ground. Being given the supremacy as the beloved, Prince Yam-Nahar usurps the power and in order to guarantee his position he sends his envoys to the assembly of the gods "in the midst of Mount of Lala". There "the gods were sitting to eat, the holy ones to dine", Baal attending upon El. Yam-Nahar instructs his envoys "at the feet of El fall not down, prostrate you not to the assembled body" and indeed proudly standing, the envoys say their speech, fire, burning fire, doth flash; a whetted sword are their eyes. They say to Bull his father El: "message of Yam your lord, of your master Judge Nahar. Surrender the god with a following, him whom the multitudes worship. Give Baal to me, to lord over, Dagon's son whose spoil I will possess".

El surrenders to this demand; "Bull his father El: Thy slave is Baal O Yam thy slave is Baal forever, Dagon's son is thy captive".

Baal, however, does not accept this verdict, at first he wants to strike the envoys but Ashtoreth convinces him not to act against an envoy. Baal, however, then receives from Kothar-Wa-Khasis two bludgeons, one named "the driver" the other "the chaser", and with their help he smashes Yam-Nahar, who at the end says twice "I am dying, Baal will reign".

Baal's reign is not yet ensured, he has another encounter with Mot (Death) and is swallowed by him. When Baal is dead, the "Lady Asherah of the Sea" tries to put her son Yadi Yalkhan on the throne of Baal, but her suggestion is opposed by El, who claims that Yadi-Yalkhan is "too weakly, he can't race with Baal, throw javelin with Dagon's son". Lady Asherah of the Sea suggests that Ashtar the tyrant be Baal's successor, but when Ashtar sits on Baal's throne in the Fastness of Zafon (north, also hidden place) his feet "reach not down to the footstool, nor does his head reach up to the top".

As Baal is dead, "life's breath was wanting among man, life's breath among earth's masses". Anath the maid, Baal's sister and consort, whose heart is for Baal "as the heart of a cow for her calf, and the heart of a ewe for her lamb" seizes the godly Mot "with sword she doth cleave him, with fan she doth winnow him, with fire she doth burn him, with hand-mill she grinds him, in the field she doth sow him, birds eat his remnants".

While Baal is dead "parched is the furrow of Soil-O-Shapsh, parched is El's soils furrow, Baal neglects the furrow of his tillage".

Somehow Baal comes back to life. "And behold alive is Puissant Baal, and behold Existent the Prince Lord of Earth. In a dream O Kindly El Benign, in a vision Creator of Creatures, the heavens fat did rain, the streams flow with honey, so I know that alive was Puissant Baal, Existent the Prince Lord of Earth". As Baal comes back to life and sits on the throne he seizes the sons of Asherah and strikes them down and after some more fights with Mot he sits on his throne, to eat the bread of honor, and drink the wine of favor. After more fights with Mot, Baal remains the Lord of Mankind. Anath, the sister of Baal also fights many other wars with dragons and monsters, including Yam (sea) Tanin, and Lotan (Leviathan) and a seven-headed dragon (Shalyat).

The adventures of Baal and Anath as told here in detail, combined with our knowledge of the people's dependence on the sources of water, can, in the opinion of the author, help in deciphering the story of the evolution of the faith of the people of the Fertile Crescent.

In the first place, as already mentioned, the Canaanite mythology includes the same themes as one finds in that of Mesopotamia and archaic Greece: the replacement of the primeval creator, either Anu, Cronos, or El, by a younger and vigorous god Illil, Marduk, Adad, Zeus, or Baal; the goddess of fertility, sister and consort to the young god, is found in all these religions. Her names change, but the task and acts remain similar whether it is Anath, Astarte, Asherah in Canaan or Inuini-Ishtar in Mesopotamia, Aphrodite or Athena in Greece. There also exists a correspondence in the deities of the second echelon, although none of them maintains the same level of importance. Thus the Moon-God-goddess, Yerach, as well as the sun-god, Shemesh, had a higher position in archaic Canaanite religion, mainly the third millennia B.C. as can be seen from the many names of ancient towns such as Jericho, Bet Yerah, and Bet Shemesh. The celestial deities declined during the second millennia in Phoenicia. On the other hand, we find these deities prevalent in Akkad and in Sumer, which had become Semitic, during the second and first millennia B.C.

Does the Canaanite mythology represent just a mixture of all the other mythologies, as many archeologists maintain, or does it preserve the roots of an authentic archaic religion?

The foregoing chapters on the evolution of the agricultural societies in the Fertile Crescent, which portrayed their great dependence on the sources of water and on the forces of nature, give a key to answering this question. One has only to take into consideration that religion, as any other social system, is bi-directional; namely, that it influences and is influenced at the same time. Another important factor one needs to take into account is that the forces of nature and the deities representing them, as well as the societies or nations and the deities protecting them, were thought to be one in pagan societies. Thus, if a certain society from a region where a certain force of nature was predominant defeated and overwhelmed another society whose local gods represented the local forces of nature, the deities of the newly evolving society would go through a period of renaming, changes in the hierarchy of their Pantheon, transmission of powers, change of cults, and exchange of authorities from one god to another. The outcome may look mixed up, but, keeping track of the relationship between environment and deity, as well as of the relative age of the different cultures, the evolutionary trend becomes apparent.

Remembering that the first agricultural societies in the Neolithic period started near the outlets of springs, it is not astonishing to note that El, the primeval deity, lives at the sources of rivers in the midst of the outlet of two deeps (Teomotaim). The spring, flowing from the depths, and giving rise to the rivers, is also the source of life. Teomotaim, Tehom, Tiamat may thus be the most ancient deity. It might be that its origin is even pre-Neolithic, namely, Upper Paleolithic, which produced the earliest sculptures of the goddess of fertility yet found.

The name of El may have its roots in the name of the ram, the symbol of fertility and power in the societies dependent on domesticated sheep and goats. Remembering the Chalcolithic temple at the spring of Ein Gedi, as well as the Chalcolithic treasure of the bronze ram heads found in a cave nearby, it may well be that the deity which symbolized the ram and that which symbolized the spring water were blended together into one, after the archaic first stage of the war between the male god and Tiamat was settled. It is interesting to note that Ein Gedi means "the spring of the kid" and that the god Pan of the Greeks, who was also the deity of springs was also portrayed as half man and half he-goat. The second title of El, Father Bull, reminds one of the Anatoliaen civilization of Catal Uyuk where the bull played the main role in the Pantheon. The bull also played an important role in the archaic civilization of Crete .

Thus, it might be a possibility that the primordial god El of the people living from flocks of sheep and goats, who also practiced irrigated agriculture along the beds of rivers fed from springs, was transformed into a complex deity comprising also the bull. This may have happened as a result of the invasion or migration of people from the north. This god was later replaced by a new god, Baal or Hadad, representing the culture of the plain dwellers. These people did not depend so much on the springs as a source of water, but on the rain. They were most probably also northern people who moved southward as the climate became more humid and thus the semi-arid regions became fit for their way of life. A more humid climate meant also the swelling of the springs and the flooding of the river basins. The correlation between rainstorm and destructive floods was eventually explained as a contest between Yam-Nahar (the ancient meaning of "Yam" was any large area covered by water, either a lake, sea or water reservoir), the favorites of the old god, and the new claimant for supremacy, Baal. As the new deity won the fight, he was acclaimed the new lord of the Pantheon. This contest could have taken place in the Jordan Valley or in one of the valleys of the Zagros, Tauros, or Pontus. The Syrian Rift Valley and especially the section stretching from Mount Hermon or Sirion to the Dead Sea seems to answer most of the general descriptions found in the ancient tablets, but at the same time it could have happened in any other similar valley with a spring issuing from the mountains at the head of the valley. The fact that the Akkadian mythology speaks of Tiamat as the opposing deity, while in the Ugaritic story, Yam-Nahar and Mot are the opponents, while El already sits on Teomotaim, strengthens the assumption presented above, that the war with Tiamat is the most archaic of the stories. It was already in existence when the war between Baal and Yam-Nahar and Mot took place. Thus, in the author's opinion, the relations between the society and the water resource on which it was dependent, as found in the Canaanite myths, indicates a change from a society dependent on water resources issuing as springs, to that of rains; and from a society for which the ram was a symbol, to that of the bull.

All these changes may be explained in two ways: either the society moved from one place to the other and was influenced by different cultures characteristic of different regions or a change in climate also brought intrusion of other societies, ways of life and idols. The fact that we are dealing here with a society, the lands of which are bordered to the south by the desert, the country of sheep and goats, and to the north by the more humid belt, the country of cows and bulls, speaks for a change of climate from drier to more humid; in other words, the migration southward of the belt of humid climate and in its wake the migration of societies and their herds of cattle and protecting deities.

In conclusion, it can be said that the survey of the evolution of the settlements of prehistoric man in the Fertile Crescent, the findings of cult-objects in their layers, and the study of their myths, gives a clue to the evolution of their faiths and fates.

It seems that during the Upper and Epipaleolithic period, namely, at the dawn of the Holocene, the main cult object was the female fertility goddess who to these hunting and collecting societies symbolized the blessing of multiplication and thus strength of family and tribe. At the beginning of the Neolithic period with the gradual settlement of the valleys fed by spring water, this goddess may have symbolized the goddess of the depths and Primeval Waters, namely, Tiamat. A gradual adoption of other deities may have been taking place during the Neolithic age. It might be that the main cult was that of the spirits of their ancestors as portrayed by the plastered skulls found in Jericho in the layers of the PPNA. The worshiping of the moon or sun or both can also be considered. With the period of the PPNB, as was explained above, a more humid climate also began that brought about the expansion of settlements outside the valley bed, due to the domestication of sheep and goat.

The people of the PPNB period also had a new cult, plus new tools and a new economic system. These people also had a building which contained a standing stone as a cult object. Was this the first symbol of El? The phallic association of the standing stone and the later associations of the use of the Mazaba in connection with the cults of the Semites may suggest that indeed El as the symbol of fertility, power, and abundance started to play the role, which until then belonged to the Goddess of Fertility.

As seen from the anthropological study in Beer Sheva, the Chalcolithic and more humid period brought with it the immigration of people from the north; and with it, to the southern part of the Fertile Crescent came the worship of the bull. The idol of the bull carrying two butter churns and the idols of the rams, found hidden in a cave nearby, suggest either integration of the cult of El as the ram-god with that of the bull, or the replacement of El, the ram-god, by El-Father Bull. In this case, the bronze rams, which were hidden in a cave near Ein Gedi, may represent the victory of the bull worshippers over the ram admirers.

The transition to the cult of Baal seems to have taken place at the end of the Chalcolithic period when a drier period started (following a series of floods that destroyed many settlements in the plains of the big rivers). The people of the rain-fed agricultural communities encroached on the river-fed plains, searching for safety, and bringing with them from their plains their god of rain, lightning and thunder.

The Canaanite or western Semitic civilization in the Fertile Crescent flourished during the Bronze Age, starting about 5200 B.P.,(3200 B.C.) and ending about 3200 B.P.(1200 B.C.)

The Early Bronze Age was most probably a humid period, with a short dry spell around 4600 B.P. (2600 B.C.). Up to this dry spell, large fortified cities were established in the arid Negev of Israel, like the town of Arad [4]. A clue to this can be seen in the previously mentioned core samples taken from the bottom of the Sea of Galilee (Fig. 2.3), especially in the curve which expresses the pollen assemblage and isotope abundance in a statistical method which eliminates the short-term fluctuations. Thus, around 7200 B.P. (5200 B.C.) when the Early Bronze period starts, there is a depletion in the isotope composition; later at about 4600 B.P. there is an increase, apparently a warm phase (the city of Arad was abandoned at that time). After about 200 years there is again a depletion in the isotopes of ^{18}O and ^{13}C; this time it coincides with a peak in pollen of olives and a low in oak. This means that a colder, more humid, period encouraged the people to cut the natural forest and plant olive in its place. At the end of this period, namely, the end of the third millennium, around 4000 B.P. (2000 B.C., Middle Bronze Age) a phase of desiccation came over the Fertile Crescent. This caused many disturbances throughout the region. The most prominent was that already described - the immigration of the "People of the West", the Amorites, into the fertile valley of Mesopotamia. Hadad, the god of rain, the equivalent of Baal (called also Baal-Hadad), of the more western Semites or Canaanites replaced the Sumerian gods, while Semitic kings named Sumuel and Avisar sat on the throne of Larsa and Isin.

As previously mentioned, among the officials in Larsa is one called Abram, and among the tribes reported to be raiding the borders of Larsa one finds the Benjaminites.

It seems that not all the tribes which were driven from the desert towards the sown lands settled down to adopt a new way of life. Some tribes, such as the Benjaminites (Sons-of-the-South, [4]) continued to live on the border between the sown land and the desert. These tribes most probably maintained their basic source of income: the herds of sheep and goats, and maybe even some cattle. The herds grazed either on fallow land (Gan Edin) or on lands beyond that irrigated by the canals, fed only by the scarce rains. It seems quite probable that, as the climate became dryer, many of these tribes could not find enough food for their herds on the margins of the southern part of the valley and, as the irrigated land was already occupied and protected by earlier tribes, they had to move northward. This might also be the reason for the travels, related in Genesis, of the ancestors of the Patriarchs from Ur in Sumer to Haran in southern Anatolia. A change of climate, or overgrazing and overpopulation caused them later to look for a future in the Land of Canaan. The route southward most probably took them along the eastern flanks of the mountainous backbone of the Land of Canaan. This area lies on the border between the forests which spread along the western flanks of the mountains, facing the rain, and the semi-desert on their eastern flanks which are in the shadow of the rain. This is an area in which, during winter and spring, one can find food for one's herds even if the year is relatively dry. But also the resources in Canaan failed and there was also a shift of population from the deserts to the less arid parts at that time, as one can learn from the results of the archeological investigations in the Negev [5,6]. (In these references the reader will find only the description of the movement of the population. As was mentioned already most archeologists did not regard aridization as a factor in such movements).

The wanderers thus had no reason to stay in the Land of Canaan, they continued their voyage south and southwest to the Land of Egypt. The reason for finding food in Egypt while Canaan suffered from hunger was mentioned in Chapter 1, and will also be discussed later when the story of Egypt is told. For the time being, it can be said that the preference of Egypt, as a land of rescue, was a function of the two different climatic regimes, one opposing the other, which characterized Canaan and Egypt. Thus, while the first went through a spell of drought, the countries from which the Nile emerged enjoyed a humid period.

According to the Bible, the patriarch Abraham was at the head of the tribe which adopted Yahawe as its only god, started to practice circumcision as an identifying sign, and did not assimilate with the local inhabitants. The question whether Abraham was a historical figure or just a prototype, namely a personification of many tales and traditions, is beyond the scope of the present study. The reader interested is referred to the general bibliography of this chapter. In the forthcoming paragraphs, the environmental hydrological issues as expressed in these stories will be discussed.

We are told that after a short stay in Egypt, the Patriarch Abraham and his dependent, Lot, came back to the Negev (Fig. 6.1), where "their substance was great, so that they could not dwell together" (Genesis 13:6). Lot decided to descend to the plain of Jordan and pitch his tent toward Sodom. This plain was "well watered everywhere, before the Lord destroyed Sodom and Gomorrah, even as the Garden of the Lord, like the Land of Egypt, as thou goest unto Zoar" (Genesis 13:10). Water is abundant there even today; however, it is wasted, flowing into saline marshes where jungles of reeds, tamarisk, and wild palm trees grow. Deep groundwater aquifers, in addition to the water flowing in the shallow alluvial layers of the Arava Rift Valley, are the source for all the springs feeding these marshes.

The Sodom plain forms the drainage basin for these aquifers. The springs emerge along fault lines on the western and eastern margins of the plain. This can be seen clearly from the satellite image (Fig. 4.1). The water is slightly brackish on the western margin and fresh on the eastern bank. Part of the water is paleowater as it was recharged tens of thousands of years ago from outcrops in central Sinai and Trans-Jordan. The present outflow is due to the tremendous storage that accumulated during those millennia.

Fig. 6.2. Mount Sodom salt plug, viewed from the east (Photo by the author)

Along the fault lines bordering the Dead Sea, hot mineral springs emerge, many of them rich in H_2S (sulfur bi-hydride), and the stench of this gas marks the areas where the springs emerge. The source of these saline springs is in deep-lying aquifers containing very saline brine. Their emergence or submergence is a function of a hydraulic balance between the overlying brackish water and the underlying brine-containing aquifers. In the vicinity of the salt plug of Mount Sodom (Fig. 6.2), brines containing CH_4 (methane) gas have been located in a deep well. It can be assumed that a severe earthquake might affect the fault line zones bordering the rift valley and open a better outlet for the deep lying brines, thus affecting the regional hydraulic balance. Saline hot water would emerge and the springs of freshwater could become salty or even hot and salty. New thermomineral springs,

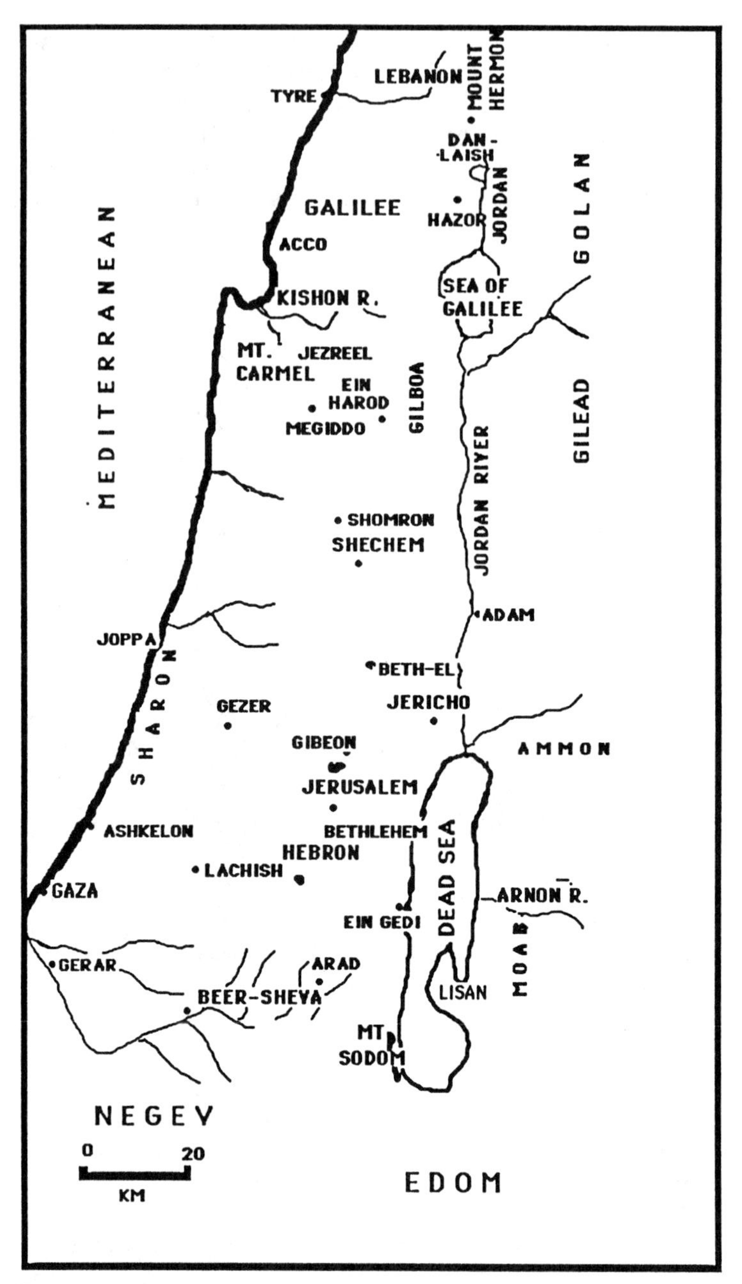

Fig. 6.1. Map of Canaan

with vapors of H_2S, might also emerge. The emergence of CH_4, methane gas, causing fire might also occur. It is suggested that the biblical verse, "then the Lord rained upon Sodom and upon Gomorrah brimstone and fire... And He overthrew those cities, and all the plain and all the inhabitants of the cities and that which grew upon the ground" (Genesis 19:24), could be an echo of a severe earthquake which caused landslides and stones to fall from the precipitous cliffs towering above the rift valley. It could also have given rise to the emergence of thermomineral springs, which resulted in the salinization of the groves and orchards, and H_2S and methane gas, causing fires and suffocation.

Before proceeding to the illumination of the stories of the wanderings of the patriarches, and the hydrogeological phenomena which may have occurred in this region, the reader is again reminded that the author does not intend to explain particular events or discuss their morals. Thus, one cannot claim that the geological and hydrological conditions of the southern part of the rift valley prove that the story of the destruction of the cities in the Valley of Sodom happened in the time of the patriarch Abraham, the way it was narrated in the Bible. However, it is claimed that, knowing the natural environmental conditions and the cultural background of this group of wanderers, one finds a correlation between these stories and this background. The conclusion of the author is, that, during the entrance of the Hebrew tribes into the Land of Canaan a severe earthquake did destroy some towns and villages in this region. The moral code of the ancient Hebrews explained the destruction in a moral context and not, for example, as the failure to sacrifice fat enough sheep to the gods.

Some of the stories about the patriarches happen in the Desert of the Negev in the area surrounding Beer Sheva (Fig. 6.1), already mentioned. The ancient Tel of Beer Sheva (Fig. 6.3) is situated near the bank of an ephemeral river-bed (wadi) draining the southern part of the Judean Mountains and the plains of the northern Negev to the Mediterranean. The wells along the river-bed of Beer Sheva are described in two instances: one, on the occasion of a covenant between Abraham and Abimelech (Genesis 21:33); and the other, when the servants of Isaac dug a well after another covenant with Abimelech, King of the Philistines (Genesis 26:19).

The story of the search and quarrels for wells of water in the river-beds of the plain of Beer Sheva is of special interest in the light of hydrogeological conditions of the northern Negev. This entire region from the foothills of the Judean Mountains down to the coastal plain was, according to the Bible, the Land of Abimelech. (A Semitic name which shows that calling him King of the Philistines is a later addition. The Philistines, who are of Indo-Arian origin, settled in this region at the time of the conquest of Canaan by the Hebrews at ca. 1200 B.C). This whole region is built of impermeable chalks overlain by impermeable loess (silt). In the river-beds permeable sand and gravel layers are found. The river-beds of Beer Sheva and Gerar also have a thick sand layer in their subsurfaces, reminiscent of the intrusion of the sea during the Neogene [7]. Due to these permeable layers of gravel and sand, and the impermeable layers of chalks underlying them, a perched water table is created along the river-beds (Fig. 6.4). Digging into the river-bed, however, does not necessarily lead to finding water, as sometimes the gravel and sands may be replaced by the impermeable loess which was washed into the river-bed and deposited at some prehistoric stage of denudation and refilling of the sands and gravel. In some places saline water is found in the sands of the ancient Neogene Sea. Thus, not only has

Fig. 6.3. Tel Beer Sheva on the bank of wadi Beer Sheva. (Air photo, courtesy of Irit Zaharoni, Bamachne Journal, IDF)

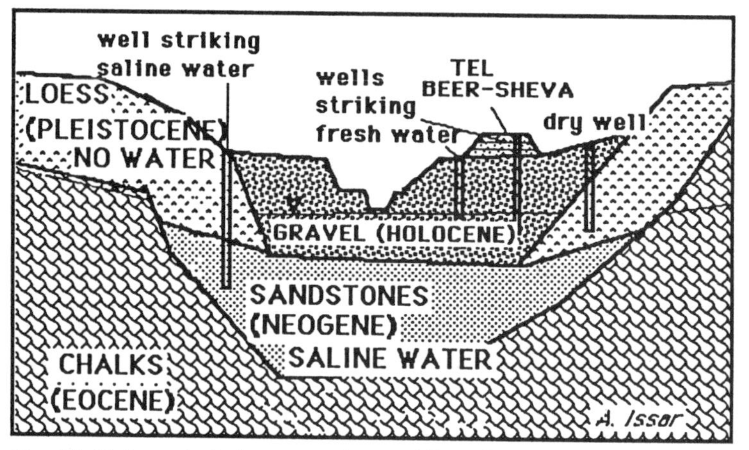

Fig. 6.4. Hydrogeological cross section wadi Beer-Sheva

one to dig into the river-bed in order to find water, as the servants of Isaac did - "and Isaac's servants dug in the valley and found a well" (Genesis 26:19) - but one has to be lucky to remain in the layer containing the "springing" water and not penetrate into the sands containing saline water. The search for water, the rivalry between the Philistines and Abraham's and Isaac's servants, the journey to the land of the Philistines in a year of drought when one had to leave the more arid eastern lands and travel along the coastal plain where rain was more abundant, all give a realistic picture of the climatological and hydrological conditions of the Negev, the land on the border of the desert.

7 A Land Like a Garden of Herbs

> The Land of Egypt...where thou sowedst thy seed and wateredst it with thy foot as a garden of herbs.
> But the land whither ye go to possess it, is a land of hills and valleys and drinketh water of the rain of heaven. (Deut. 11:10,11)

The aridization phase, mentioned already, which affected the region at about 4000 B.P. (2000 B.C.) at the end of the Early Bronze Age, was characterized by the desertion of perennial settlement sites and the increase in the number of sites of nomads on the margins of the more humid regions (Chap. 6 [5,6,7]). The wave of nomads who were not satisfied with the meager resources the desert could provide, and did not have the courage and power to forcibly take the goods of the irrigated lands, moved along the borders of these lands trying to get something from the richness of these countries in exchange either for their products or for their services. In this context, one understands the many witnesses to the migration of Semitic tribes from the Fertile Crescent to Egypt, one of which is the famous wall picture from an Egyptian tomb in Beni-Hassan painted around 3900 B.P. (1900 B.C.). The picture portrays a group of people of Semitic origin who are about to enter Egypt; the leader, Avisa, dressed in a colorful garment, makes a sign of obedience to the Egyptian governor with one hand, while in his other hand he holds a he-goat, either as a present or as a indication of his source of income (similar to the bellows portrayed on the back of the asses). The two Egyptian officials accompanying the group of immigrants suggest that the entrance of the Semites was a move acknowledged by the Egyptian government.

The migration of the Patriarchs to Egypt after they reached the Land of Canaan can thus be understood in the general framework of the movements of Semitic tribes from the semi-arid Fertile Crescent to its southern edge. Although this area is more arid it is supplied by the perennial water source of the river Nile

It is thus interesting to note the fact that the two valleys, Mesopotamia as well as Egypt, each of which is more arid than most of the lands forming the Fertile Crescent, could serve as a firm basis to agricultural civilizations, and even supply a source of income to the refugees. This, as previously explained, is due to the fact that the two regions, Mesopotamia and Egypt, are transversed by big rivers, the sources of which are from rains falling on more humid terrain. Yet the stability of supply is a result not only of the abundance of rain in the drainage basin of the supplying rivers, but also of the storage capacity which is part of these river systems. In Mesopotamia, the storage is supplied by the snows, which melt in spring, thus spreading the period of supply into the early summer, and subterranean space of the aquifers of the Anatolian and Zagros mountains. In the Nile system, the regulating storage is supplied by the lakes and marshes of subequatorial Eastern Africa. Another difference between the Mesopotamian and Egyptian river systems is the climatic belts to which they belong. While the Mesopotamian system belongs to the Mediterranean climatic belt, the Nile is fed by the tropical rains falling on east central Africa and the subtropical monsoonal system stretching north to the equator. The latter system is seasonally, as well as periodically, in an off-phase regime to that of the Mediterranean, as was explained in Chapters 1 and 2.

The Nile is the longest river in the world, measuring about 6600 km from its headwaters in Rwanda to its outlet in the Mediterranean Sea in Egypt (Fig. 7.1). Ever since man started his research into the mysteries of nature, the sources and regime of the Nile have posed an enigma. The Egyptians considered the Nile as one of their deities and saw in its perennial flow and seasonal fluctuation a mystery which should not be explained but worshipped. People with more inquisitive minds looked for a more natural answer, and one of these was Herodotus, a Greek who lived between 490 to 425 B.C. In his book The Histories [1], he writes about "why the Nile behaves precisely as it does. I could get no information from the priests or anyone else. What I particularly wished to know was why the water begins to rise at the summer solstice, continues to do so for a hundred days and then falls again at the end of that period, so that it remains low throughout the winter until the summer solstice comes round again in the following year". After failing to obtain a logical answer from the Egyptian priests, who most probably regarded this inquiry as blasphemy, and after arguing against the answers given by "certain Greeks hoping to advertise how clever they are", Herodotus gives his opinion. He attributes the fluctuations in the flowing of the Nile to seasonal difference between the lands on the shore of the Mediterranean which he knew and the lands south of Egypt from which waters of the Nile came. In his words, "...during winter the sun is driven out of his course by storms towards the upper parts of Libya. It stands to reason that the country nearest to and most directly under the sun should be most short of water..." When the rough winter weather is over, the sun resumes its normal course in mid-heaven and from then on exercises an equal attraction over all rivers (namely, evaporates them) in winter, then all rivers but the Nile run in flood, because a great volume of rainwater is added to their volume. The Nile, on the other hand, behaves in the opposite way. In a preceding chapter, after proving that the soils of Egypt are alluvial and were brought and deposited by the Nile, Herodotus states, "It seems to me therefore that if the land continues to increase at the same rate in height and extent, the Egyptians who live below Lake Moeris in the Delta and thereabouts will, if the Nile fails to flood, suffer permanently the same fate as they said would some day overtake the Greeks for when they learned that all Greece is watered by rain and not as Egypt is by the flooding of rivers, they remarked that the day would come when the Greeks would be sadly disappointed and starve."

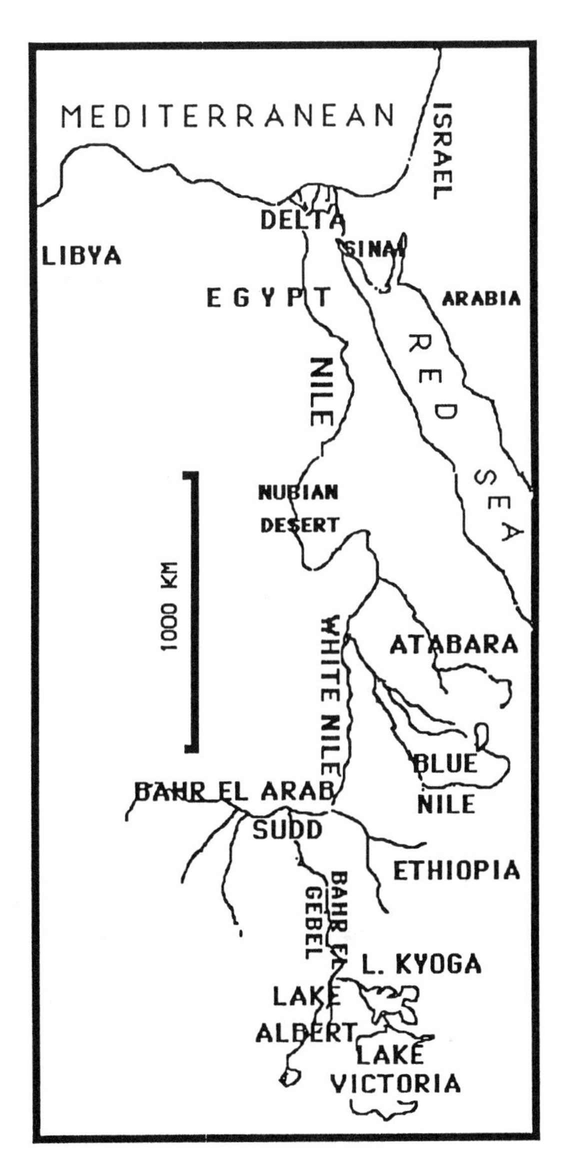

Fig. 7.1. The Nile and its tributaries

In this comment Herodotus has described the difference between the two systems of agriculture, the river-dependent system and the rain-dependent one. As already explained, the difference is the long-term storage capacity, which enables the big rivers to flow permanently, although fluctuations which are a function of seasonal variations cause seasonal changes in the flow of the river. We also know that changes of climate may also cause a change in the rate of flow, which may cause catastrophes when the changes are extreme, by either flooding or by diminishing to famine level.

The secret behind the relatively high constancy of the flow of the Nile, as well as its periodical, accurate fluctuations, lies in the combination of several factors. The first is the climate, which is dependent on the tropical system of the Trade Winds and not on the Westerlies which activate the Mediterranean system. For the explanation of this factor, which is also important for understanding other chapters in this book, some climatic background is needed.

The Nile is fed by rain falling on the subequatorial countries of East Africa, the source of moisture being southwesterly air streams from the Indian Ocean and the equator. The oceanic air blows inland in the summer, causing all the rivers north of the equator to flow. These summer rains affect mainly the Blue Nile and the Atabara rivers which originate from the north and central highlands of Ethiopia (Fig. 7.2). Thus, the floods of the Nile start in June and reach their peak in August. The White Nile, on the other hand, is fed by rains falling on equatorial Uganda and southern Ethiopia, where there are two rainy seasons, thus providing more evenly distributed runoff. Another very important storage and regulating factor is the gigantic swamps of the Sudd into which the White Nile flows, and which cause the suspended silt of this river to be deposited and the supply of water to be regulated. In terms of relative contributions, the Blue Nile contributes about 60% of the average annual water supply, the White Nile about 30%, and the Atabara about 10%.

Egypt, from the northern foothills of the Ethiopian highlands to the shores of the Mediterranean, lies in the desert belt girdling the subequatorial monsoonal-trade-wind-affected belt to the north.

The desert conditions are a result of the descending air masses which rise full of humidity over the equator, causing the heavy typical rains characterizing the equatorial belt. The descent of the air masses causes them to become hot and dry.

The annual precipitation over most of Egypt is less than 50 mm and only along the coast does it reach 100 mm. The people of Egypt depend almost entirely on the Nile and it is no wonder that they regarded this river as a deity. The river itself was called Item and its spirit was a deity called Hapy portrayed by a man with big breasts and a clump of papyrus on his head. The Nile supplied not only water but also fertilizer as the floods of the Blue Nile and Atabara brought large quantities of silt which were spread over the fields by the flood waters. These black silts brought by the Nile were called "Keme" (The Black), while the silts brought from time to time by desert floods generated by Mediterranean storms were red and this desert silt was called "Deshret" (The Red). As we will see later, this difference in colors sheds some light on events recorded in the Bible.

The deity of the Nile was only one among 80 other deities which were worshipped by the ancient Egyptians. In the present study, we are primarily interested in the deities that are connected with the realm of water and enable us to understand the role of this realm on the history of the people living near to and dependent on this water.

It is thus no wonder that the river was deified. The wonder is that it is not one of the most important gods for Egypt. The reason for this may be that when the people moved into the Nile Valley they brought their chief gods with them. As these primeval gods symbolized the powers of nature in their lands of origin, they simply added to their Pantheon the local divine powers which became important only after they settled in the valley of the Nile.

Who were the people of Egypt and from where did they come? The most ancient artifact made by man in the earliest Stone Age came from eastern Africa. In the now arid parts of southern Ethiopia and northern Tanzania evidence has been found of the existence of a hominid tool producer more than a million years ago. This hominid went through several evolutionary steps, starting in the most Lower Paleolithic, continuing through the most Upper Paleolithic and into the Epipaleolithic period, which lasted until about 10,000 years B.P. During the last Ice Age (ca. 80,000 to 10,000 B.P.) the deserts of north Africa became more humid, which allowed a large population of hunters to spread over their plains, hunting the many animals which lived in the savanna-type environments that characterized many depressions in this vast country.

At that time, subequatorial Africa, south of the Sahara, became more arid, as can be seen from the low level of the lakes of this part of Africa.

The end of the last Ice Age brought about the aridization of the Saharan belt, so that the people who roamed the wide savanna-type plains which became arid had to search for a refuge. They migrated either south to subequatorial Africa, which became humid, or to the desert oases and the valleys, watered by either springs or rivers. Remains from the Neolithic period are rather rare in Egypt. Some Late Neolithic relics were found in the Faiyum area. The emmer, barley and flax found there disclose connections with Asia, but the tools and weapons clearly indicate African traditions. During the Chalcolithic period man had already settled at many places along the river. Various sites containing relics of typical cultural traditions have been excavated. The people of this period already believed in life after death, as they provided their dead with a great supply of objects to help them in their new form of "existence." Cultural connections with other parts of the Middle East can be asserted from the female figurines with especially emphasized procreative attributes, symbols of a fertility cult.

While the bulk of the people of Egypt were the descendants of the prehistoric tribes that settled along the Nile when the desert areas first became dry, the ruling classes might have been conquerors from abroad, most probably from Mesopotamia.

At about 5400 B.P. (3400 B.C.), a drastic cultural change affected Egypt. The country was divided into two kingdoms, one of Upper Egypt, the other of the Delta region of Lower Egypt. At the same time the art of writing appeared as well as other arts typical for Egypt, such as their architecture, drawings and sculpture.

Various paintings of this period show a Mesopotamian influence. An excavated knife handle shows a sea battle between Egyptian ships and others which look Mesopotamian. Findings in graves show that people with a different anatomy (larger, with a differently shaped skull from the ordinary Egyptian), and using special funerary architecture, formed some kind of ruling aristocracy. These people were named also "Followers of Horus". They were divided into ruling houses, each governing one of the kingdoms of Egypt.

At about 5200 B.P. (3200 B.C.), the two kingdoms were united into one by the King of Upper Egypt, Menes or Nar-mer. He is believed to have founded the town of Memphis, the original name of which was Het-Kau-Ptah, the House of Ptah, which later was transformed into Aiguptos, hence Egypt. With King Nar-mer starts the First Dynasty which, with the Second Dynasty, continued until 2686 B.C. During this period, also called the Archaic Period, the art of writing developed to the high level of hieroglyphics found in the pyramids.

The religion which developed at this time was the amalgamation of the native cults of the local tribes and those imported by the conquerors. The gods of the conquered were not suppressed but were annexed to the Pantheon, or even adopted as sons of the conquering deity. They remained chief gods in their original provinces, to be worshipped for millennia.

At the time of the unification, the rulers worshipped the sky god Horus, whose symbol was the falcon. Seth was the god of the original people. The replacement of Seth by Horus after a period of struggle is narrated in mythological tales of the triumph of Horus, symbolizing good, over Seth, representing evil.

During the second dynasty there came a fusion between the sky god Horus and the sun god Re and the king identified with Horus became the son of Re. A local god in the Delta area was Ptah, who was believed to be the creator of the universe. Another important deity originating in Lower Egypt was Osiris. He was believed to have been killed by Seth and avenged by his son Horus. The Pharaohs were worshipped as demi-gods, believed to be the descendants of Horus. This most probably symbolized the victory of the god of the conquering race over the local deity.

The worship of many deities which sometimes overlapped each other in their dominions characterized the ancient Egyptian religion and, in some ways, can be compared to the Hinduistic philosophy of religion which tolerates different answers to the same problem as different facets to the same phenomenon. For instance, the creation of the world was attributed at Memphis to Ptah. At Heliopolis, Re the sun god was believed to be the creator. At Elephantine, it was believed that Chnum, the god in the form of a ram, was the creator [1, 2, 3, 4].

The ancient controversy between the ram and the bull worshippers was solved by the Egyptians adopting also the bull, Apis, which symbolized strength in war and fertility. He was believed to be the manifestation of the god Ptah, the creator. His cult on a national basis was established most probably by Menes, the founder of the First Dynasty. Hathor was the sacred cow, patron of the skies, love, and joy.

Were these gods remnants of the prehistorical cults of the people of the rams and the bulls? Most probably, yes. The Nile valley seems to have been a huge religious 'soup' to which every ancient cult became an added ingredient, but not the main component.

It is thus interesting to inquire whether the people coming from Mesopotamia at the end of the Chalcolithic period to seize power and create the ruling class also brought with them their cults from the land. We heard that they were the people of Horus the hawk, a cult which was not recorded as such in the ancient Near East. On the other hand, we know that, since the First Dynasty, the worshipping of the sun had started to become a national religion. Horus, too, was worshipped as the god of the sky, and thus goodness, in contrast to Seth, who was most probably the god of the aborigines of the Valley of the Nile. He was represented by some unidentified animal with the appearance of a dog and the head of an ant eater.

It thus seems probable that the people who came from Mesopotamia or from another part of the Fertile Crescent were worshippers of the sun and, although they accepted the local gods as members in the club of the deities, they insisted on their being second class members and even threw out the former head, Seth, denouncing him as a symbol of evil.

From Egyptian mythology one can discover what could have been the state of the religions of the Chalcolithic tribes that invaded Egypt. We saw that they worshipped the sun, the hawk, and the bull. In the previous chapters we saw that the cults of the Bull and the Sun were practised during the Neolithic and Chalcolithic periods in Anatolia, and even reached Canaan. One can thus assume that some time during the Chalcolithic period the climatic conditions enabled the people from the north to reach the border of Egypt. When the climatic conditions worsened at the end of this period, they looked for a refuge for themselves, their cattle, and their gods in the land with a perennial supply of water.

It can also be claimed that these people did not come from one of the well-established, settled Mesopotamian cultures, as one can find references in the Egyptian mythology analogies only to rudimentary items characterizing the Mesopotamian religions. As will be shown later, most beliefs are lacking the story of the Deluge and if similar myths exist they are a very distant echo of something long forgotten.

Yet one can find in the Egyptian stories of creation some metaphysical ideas which are related to those found in the book of Genesis and the Mesopotamian myths of creation. The most prevalent is that of the Primeval Waters as the primary entity from which all the world was formed. This idea is common to all the Egyptian accounts of the origin of the universe. Every creation myth in the different Egyptian mythologies assumes that, before the beginning of things, the Primordial Abyss of waters was everywhere, stretching endlessly in all directions [6]. The rising of the Creator from the Primeval Waters was also the manifestation of the creation of land, light, order, and life.

The "precreation" state, namely, when Atum was still immersed in the Primeval Waters, was a state of unhappiness and pain. He was very weary and inert; thus the waters represent helplessness and chaos. At the same time, immersion in them means going back to primeval innocence.

The Primeval Waters were represented by the deity Nun. In many texts, the description of the Primeval Waters is given in negative forms of speech, defining it by what it is not: "Where the Universal Lord dwelt when he was in the infinity, the nothingness, and the listlessness." According to the cosmogony of the cult center of Hermopolis in Central Egypt, the waters produced four beings out of themselves: (a) Nothing, (b) Inertness, (c) Infinity, and (d) Invisibility or Darkness. These were given female counterparts and were worshipped as demi- gods or goblins in the town of the "Eight Ones" or Shumunu (Eight), later called Hermopolis. They formed the Primeval Egg in the darkness of Father Nun. From this egg emerged the bird of light. Other sources maintain that the Primeval Egg coming from the Primeval Waters was filled with air which separated Geb, the earth, from Nut, the sky, a function usually attributed to Shu aided by the wind spirits. The act of separating Geb from Nut also has some vague similarity to the separation myths discussed previously, in the Mesopotamian and Biblical stories. In Memphis, the capital of the Old Kingdom, the High God worshipped was Ptah. He is described as floating in his reed boat across the sky-ocean, looking down upon his creation.

In a more abstract myth of creation, the Primeval Waters are symbolized by sleep and primordial chaos from which the living soul struggles to free itself. The High God rises from the Primeval Waters as a high hill on which Heliopolis as a cult center was built. The appearance of Atum, the creator, was as the Primeval Mound, which was also the creation of light, since the waters were in absolute darkness. Atum, or "The Complete One", was afterwards worshipped as Re, presented as the sun.

The worship of Atum-Re as High God flourished during the Old Kingdom (from ca. 2780 B.C. to ca. 2500 B.C.). This was the period in which the pyramids were built. The main religious center was at Heliopolis (today a suburb of Cairo), the priests of which gained supreme power and had their cosmogony inscribed in the pyramid texts. The idea of Atum as the only High God went through a series of changes during the period of the Old Kingdom. One of its presentations is in the form of the scarabeus beetle. The insect, observed in the act of rolling his egg enclosed in a ball of dung, symbolized Atum as he came into existence as the rising sun. The different manifestations were explained in the following way. Atum is the aboriginal invisible god who is seen in his visible forms as Khorper, the scarab, and Re, the sun in the sky. Gradually Atum, who is invisible, becomes the night sun as it journeys through the underworld.

Another manifestation of the creator is that of a serpent. In the pyramid texts, one finds the following narration:

"I am the outflow of the Primeval Flood, he who emerged from the waters.
I am the 'provider of Attributes', the serpent with its many coils".

The serpent, however, symbolizes not only the emergence of creation from the water but also the end of existence --"the great surviving serpent when all mankind has reverted to the slime". A conflict between Re, the sun god, and an enemy serpent is also encountered in the Heliopolis religion, the serpent appearing in the form of a curly-haired maiden, a legend in which one can find some relation to the story of the Snake and Eve in Paradise.

The Old Kingdom continued from the Third Dynasty, starting at ca. 2780 B.C., until the end of the Sixth Dynasty, at about 2250 B.C. The transition between the Old and the Middle Kingdoms was a period of rebellion and anarchy, yet intense literary activity flourished. In many of the remaining texts from this period, mainly the Coffin Texts, there is evidence of a more abstract way of grasping the acts of genesis and world order, but still the notion of the Primeval Waters as the beginning of everything prevails.

"I was the Primeval Waters, he who had no companion when my name came into existence" (Coffin Text 715, [5]). While still in the water, the spirit of the creator becomes a circle which is replaced by the symbol of the cosmic egg; later it turns into Hahn, which is the cosmic wind which divides and separates the sky and the earth. From the creator emerge Command and Intelligence who go round the circle of total being giving everything its name, namely, its distinctive characteristics.

"I am the Eternal Spirit, I am the sun that rose from the Primeval Waters, My soul is God, I am the Creator of the Word, Evil is my abomination, I see it not. I am the Creator of the order wherein I live. I am the Word which will never be annihilated in this my name of Soul" (Coffin Text Spell 307, [5]).

In conclusion, it is the opinion of the present author that the Egyptian myth of the Primeval Water's does not differ greatly from that of the Primeval Waters symbolizing chaos in the Mesopotamian and Biblical myths, namely, "The primeval Apsu, Mummu and Tiamat, the waters of which were mingled together" of the Babylonian-Sumerian Enuma Elish and the "Darkness over the deep" of the Hebrew Genesis. A similarity which points to a distant, yet common, origin.

Another point of resemblance between the Egyptian religion and that of the Fertile Crescent, the origins of which the author believes have a common archaic basis in the cult of the god killed and resurrected, namely of a deity which follows the cycles of the seasons: death symbolizing the dryness of the summer, and resurrection, the sprouting of the new vegetation after the rains or floods. In the Mesopotamian and Canaanite mythologies, it is the death and resurrection of Tamuz, Dummuzi, and Baal, and in Egypt, it is Osiris.

In Egypt, an extremely arid land surrounded by a desert which touches the irrigated fields, the dry season just before the floods were due had the full meaning of death if the floods of the Nile failed to arrive. The floods cover all the land turning it into a vast lake, bordered on both sides by desert. The water had to retreat before sprouting could start. All these events are symbolized in the myth of Osiris. It starts with a golden age when Osiris was the ruler of the land, and order and civilization were at their height. This was destroyed by Seth, his younger brother, who killed his brother, mutilated his body and threw it into the Nile. Some legends tell of Osiris being put in a coffin and drowned in the Nile. Isis, the sister and wife of Osiris, (compare with the Mesopotamian Ishtar, and the Canaanite Anath) helped by her sister, Nephtys, finds the body, collects all its parts, brings it to land and revives

Osiris. She bears him a son whom she hides in the swamps of the Delta. This is Horus, who grows up and takes vengeance for the evil done to his father, by attacking Seth. The war between the gods is a long episode during which order is disturbed in the world. In the end the gods intervene. An arbitration by the god Thoth decides that Horus is the legitimate heir of his father, Osiris. Horus then travels to the underworld where his father still lies, to bring him the news. Osiris revives and becomes the spirit of new life and growth. Seth is punished by being ordered to carry Osiris; he becomes the boat on which Osiris is carried on the Nile during the fertility festival.

The cult of Osiris went through many stages of evolution until, at the end, Osiris was regarded as the chief deity of the Egyptian Pantheon. The story of Osiris concerns us in its link with the myths of Mesopotamia and Canaan, in the basic theme of the fight between the primordial powers of disorder and a new power which represents order and organization. Again one can see that, although the Egyptian religion developed and evolved independently of the other religions of the Levant, yet it had in its roots some common denominators with those of Mesopotamia and Canaan.

The author suggests that these roots stem from prehistoric times, mainly the Chalcolithic period, before the Mesopotamian civilization developed its main characteristics. Thus, while one finds similarities in the basic themes of the archaic myths of creation and fertility, the story of the Deluge, which was developed in Mesopotamian civilization in the preliterate period, is very different in the two cultures. In the Egyptian mythology, this story starts with Re asking Hathor, the Cow-Goddess, to destroy mankind because they plotted against him, but when she proceeds with this task he changes his mind. He orders servant girls to prepare an intoxicating red brew from barley and red ochre with which he floods the country to the depth of three palms. When Hathor comes at dawn to slaughter what remained of mankind, who meanwhile had fled to the desert, she sees her reflection in the water. She drinks the water, which is beer, becomes drunk, and forgets her task of annihilating mankind. After that the ritual preparation of beer by servant girls for Hathor was part of the festivities during the annual festival. As can be seen, there is very little in common between the two legends. While the Mesopotamians described the flood as a disaster, in Egyptian it was described as a savior. Thus, the present author's opinion is that the story of the Egyptian Deluge is connected with a severe drought symbolized by the people fleeing to the desert. The drought was brought by Re, symbolized by the sun. The same sun also brought the silt-laden floods of the Nile which saved the people.

Thus, it might be that the Chalcolithic period (approximately 5400 B.P.), when a big change influenced Egypt to become abruptly converted from a state of advanced Neolithic village cultures to two well-organized monarchies with many signs of Mesopotamian incursion, was also the time in which the branches of the archaic myths of creation and resurrection were grafted onto the local myths to later develop into the very special Egyptian civilization.

A short summary of the history of Egypt as an introduction to the next chapter will conclude the present one.

Some time at the end of the Chalcolithic period (ca. 5200 B.P.), the two lands of Egypt were unified, and writing was developed into hieroglyphics.

The period between 3200 and 2686 B.C. is called the Early Dynastic period (Dynasty 1 and 2), during which the capital was established at Memphis which was the cult center of Ptah.

During the period of the Old Kingdom (Dyn. 3-6, from 2686 to 2181 B.C.) the first pyramids were built, beginning with step pyramids and developing into the classical pyramids at the end of this period.

During the period termed the First Intermediate Period (Dyn. 7-10, 2181 to 2050 B.C.), Egypt went through a stage of social and political strife. This helped foreign peoples, driven from their lands by famine, to penetrate and settle in Egypt.

The period of the Middle Kingdom which followed (Dyn. 11 and 12, 2050 to 1786 B.C.), was one of internal peace and prosperity. From this period connections of trade and culture with the Land of Canaan known from archeological findings in the two countries. At the end of the Middle Kingdom, foreign rulers established themselves on the throne of the Pharaohs and ruled Egypt for about 100 years.

Fig. 8.1. The Delta of the Nile as seen from space (A NASA picture)

8 The Ten Plagues - Change of Climate

> And Joseph said unto his brethren ... For these two years hath the famine been in the land and yet there are five years in which there shall neither be ploughing nor harvest ... and he made me a father to Pharaoh and lord of his house and a ruler throughout all the Land of Egypt. (Genesis 45,6-9)
>
> .... and the Lord sent thunder and hail, and the fire ran along the ground; and the Lord rained hail upon the Land of Egypt (Exodus 9,23).

In 1780 B.C., Egypt underwent an extreme change in its regime. The rulers of the country were no longer Egyptians, but a foreign people of special military talents who were called Hyksos (This word is derived from Helenistic sources; the original Egyptian name was like Haka Hashut meaning "rulers of Foreign Lands" [1]). They came from the northeast. In the Land of Canaan they overwhelmed the local inhabitants, probably by their skill in mobile warfare, using light war chariots drawn by horses onto the battle field and siege machines to conquer the towns. Afterwards, they refortified these towns with huge earthen ramparts to protect the walls against siege machines.

The exact origin of the Hyksos is not yet known. The Egyptians refer to the raiders coming from the east, as "Apiru", which reminds us of the name "Hebrews". Some of the kings of the Hyksos also have Hebrew sounding names. Opinions differ on whether the Hyksos were descended from the Hurrians, who originated in the northern part of the Fertile Crescent, probably Anatolia, or were Amorites who embraced Hurrian elements. It is believed that the movement of the Hyksos was multiracial. The Hebrews most probably followed the hard core of warring tribes who seized power in the Canaanite-ruled cities and later moved southward into Egypt.

What could have caused this migration of people? We have seen that, from ca. 4200 to about 4000 B.P. (2200 to 2000 B.C.), a period of dry years compelled the

Amorite people to escape from the parched semi-arid lands into Mesopotamia, where they replaced the ruling society of Sumer. At this time, which coincides with the First Intermediate Period in Egyptian history, we know from Egyptian sources that the infiltration of people from Canaan into Egypt had also begun. As one can see from the records, some of the newcomers to Egypt were welcomed by the Egyptian authorities probably because of their skill in some of the arts needed by the Egyptians. The picture from the tomb at Beni Hassan, described in the preceding chapter, especially of the bellows on the back of the donkey, surely tells something.

Why was there not also a penetration by force across the borders into Egypt, as happened in Sumer? Although the historical records tell us that from ca. 4200 to 4000 B.P. (2200-2000 B.C.) there were problems on the borders of Egypt, these were more with the Libyans to the west. Libya is also a Mediterranean-type arid zone, affected in the same way as the Fertile Crescent. Thus while the rainstorms, and flow of springs and rivers of Mesopotamia, Syria, Canaan, and Libya are caused by moist air originating over the seas of the Northern Atlantic and Mediterranean, the Nile is fed by rains originating over the Indian Ocean and the tropical belt. Thus an aridization of the Mediterranean climatic zone, which affects the Euphrates and Tigris, will not affect the Nile. On the contrary, it may happen that a spell of dry years, caused by the northward migration of the desert belt may have simultaneously caused the movement of the monsoonal belt northward, which would mean abundant rains over Ethiopia and the southern Sahara, and plenty of water in the Nile. This would, of course, mean abundant food in Egypt. Thus, while countries depending on water from rivers in the north, such as Sumer, Akkad, and Canaan went through an economic crisis and thus became vulnerable to the attacks by famine-stricken desert tribes, Egypt would not suffer the same economic crisis. At the same time, the Sinai Desert, separating Egypt from Canaan, became drier, making it difficult for large hordes of people to cross. Small groups could still move from one small oasis to another to reach the Egyptian border. The Egyptians, who had enough food, may have allowed these people from the desert, who could supply services needed by their rich society, to enter their country peacefully, while at the same time they fought back the tribes from Libya.

The movement of the Hyksos down from the semi-desert of the Anatolian plateau into the coastal plains of the eastern Mediterranean might thus have started with the drought that caused the great Amorite movement in about 4000 B.P. (2000 B.C) Because of the same drought, some time afterwards, the Hebrews may have started their wandering movement along the western margins of Mesopotamia which was at that time already occupied by the first wave of Amorites. Not being able or not being willing to find a permanent place to settle, they followed the migratory trend southward with the wave of Hyksos, in the hope of finding a source of livelihood in the fertile coastal plain or the valleys of Canaan.

Some of these tribes were able to secure a holding in the more fertile valleys irrigated by perennial springs. The family of Lot, the son of Haran, is an example of such settlement. In broader terms, the inhabitation of the more humid mountains of Trans-Jordan by the Semitic Ammonites, Moabites, and Edomites may reflect such a process. Thus the migration into Egypt can better be described as an act of infiltration and settlement by small groups, mainly into the Delta zone of northern Egypt (Fig. 8.1).

The biblical stories of the migration of the Patriarchs between Canaan and Egypt reveals the character of this type of movement, which reflected the changes between dry and wet spells of a relatively arid climatic period. The later settling of Jacob and his sons in Egypt most probably occurred when the climatic conditions became too arid in the Land of Canaan. From the more detailed biblical narration of these events, one hears some echoes of the environmental and socio-economic background which existed in these lands in the period of the settling of the Hebrews in Egypt. Abraham, and later Isaac, descend to Egypt and return to Canaan. Later, Isaac settles in the area of Beer Sheva, the semi-desert part of the Land of Canaan. He has several disputes with the local population who were called Philistines (after the name of inhabitants who will later occupy this region), but had Semitic names and were a part of the Canaanite tribes living in this area. We have already described the hydrological background of these disputes. Following a family feud, Isaac's son Jacob escapes to Mesopotamia, where he joins his former clan, marries, raises a family, then migrates back to the Land of Canaan. On his return through the Syrian desert, he encounters difficulties with his brother's tribe, the Edomites, but is able to reach an agreement. He then enters the Land of Canaan and moves with his tribe and herds between the fortified Canaanite towns, some of which might already have been occupied by Hyksos clans. At the town of Shechem, a feud, caused by the breaching of the code of honor by the local people, brought about the storming of the city by the sons of Jacob. To avoid vengeance, they moved quick southward to the border of the desert, which was the natural habitat of these tribes. A spell of severe drought causes the family to request permission to enter Egypt. They are helped by a member of the clan who had reached Egypt before them and risen in the local government to be the viceroy. This might have been due to the Hyksos and Apiru people's pervasion of the government of Egypt during the Intermediate Period, as surveyed in the preceding chapter.

Historical records show that at about 3700 B.P. (1700 B.C.) the Hyksos gained full control over Egypt and a King of Hyksos origin occupied the throne. It is quite logical to assume that at first officials of 'Apiru origin reached high posts in the government of Egypt, enabling their clans still living in drought-stricken Canaan to enter Egypt and settle on irrigated land in the Delta area. Most probably, these people were granted special rights and enjoyed their stay in this country.

In 1558 B.C., the Hyksos dynasty was deposed and replaced by an Egyptian king. This led, apparently, to the Apiru being deprived of their special privileges and even being enslaved and forced to work in the royal construction projects.

The downfall of the Hyksos might have been connected with a climatic change which caused a more humid phase to cover the Fertile Crescent but a dryer one to affect Egypt. Hunger and strife among the Egyptian farmers, on one hand, and the concentration of wealth and power by an alien ruling class, on the other, may have triggered the period of unrest which brought the overthrow of the hated dynasty in its wake. The revolt against the Hyksos was led by the two brothers, Kamose and Ahmose, who finally defeated the foreigners and expelled the Hyksos. After finally ascending the throne in 1558 B.C., the Egyptian King Ahmose reorganized the country and the army. He and his descendants, who comprised the 18th Dynasty, even organized war missions into the Land of Canaan. From the engraved memorials of the kings of this dynasty it can be learnt that in some way Canaan became a protectorate of Egypt.

Toward the end of this dynasty, one of the kings brought about a religious revolution by shifting the seniority in the Egyptian Pantheon from Amun to Atum or Aten, the Sun God. This king also changed his name to Akhenaten and moved the capital of Egypt southward to a place called today El-Amarna, to get away from the influence of the clergy devoted to Amun. The archives of the court containing many clay tablets written by the kings of Canaan, Akkad, and Hittie were found in this new capital. As can be seen from these tablets, most of the kings of Canaan regarded the King of Egypt as their protector and arbitrator in local feuds. The religious reform did not continue after the death of Akhenaten. The priests of Amun again gained control of the state religion, recrowned their god as chief, and removed all trace of the sun reformation.

Some scholars maintain that the biblical Exodus took place during the reign of Akhenaten, who was occupied with religious matters. Some see in the sun religion some kind of a shift toward monotheism and compare it with that introduced by Moses. All scientists who believe that the Exodus of the Bible was a historical event, place it between the end of the Hyksos rule over Egypt at the beginning of the 18th Dynasty (1558 B.C.) and the reign of Merenptah of the 19th Dynasty (1220-1238 B.C.). The later date is based on the fact that in a stele of this king the term Israel is first mentioned, claiming that it was destroyed for eternity. Although the last statement was undoubtedly premature, this document shows that the Israelis-Hebrews were considered as a nation one has to fight, destroy, and boast about. Some connect the biblical Exodus with the volcanic explosion of Mount Thera in the Aegean Sea, which is known today to have occurred at about 1650 B.C. Some connect these two events, claiming that the volcanic explosion caused a tsunami (a flood by sea waves caused by a submarine earthquake) which drowned the Egyptian army. Lately, ashes from this eruption were found in the clays of the Egyptian Delta and this finding was also related to these two events. (The various opinions regarding the date of Exodus are discussed in [2]). The date of 1650 B.C. was derived from examination of cores taken from the glaciers of Greenland and dated by ^{14}C [3,4,5]. The ice shows traces of acidity caused by this volcanic explosion. Although it is apparent that the explosion of Mount Thera had an environmental effect on the eastern Mediterranean coasts, either as flooding by waves or as ash falls, yet the author does not suggest fixing exact dates, but examining the possibility that there were anomalous environmental and historical events, which remained in the collective memory of the people, later to be connected together, in a cause and effect bond.

The way suggested in this book to connect these events is the following: A cold spell (presumably but not necessarily) was caused by the volcanic explosion. This was due to the fine ash being dispersed in the atmosphere and reducing the solar radiation reaching the earth. This caused the glaciers to expand, which caused the movement southward of the belt of the Westerlies. This may have pushed southward the intertropical arid belt, causing a reduction in the precipitation over Ethiopia and thus less water to flow in the Nile, which also meant less water and food to spare for any alien population.

Further evidence of a cold spell is from research carried out in the Libyan desert by a German team from the University of Berlin. It shows that, at about 3420 ($\pm$230) years B.P. (dated by ^{14}C), a humid spell affected this desert. On the other hand, more or less at that time (between 3500 and 4000 years B.P.), a dry spell affected Kenya [6].

The southward movement of the arid belt and, in its wake, the belt of the northern low pressure, may have had another effect on the natural environment of Egypt, like that which affected it during the Last Glacial period. Rain storms coming over the Mediterranean would penetrate southward into the deserts of Cyrenaica and Egypt, causing rain to fall on areas which for centuries had received only a few millimeters of rain. Such rain would cause severe changes in the natural habitat, as well as the land forms. The dates of the climatic changes discussed earlier were fixed by radiocarbon methods in which an inaccuracy of about 100 years is reasonable. This means that if Exodus was really connected with this climatic change, it cannot be placed exactly, but sometime at the beginning of the 18th Dynasty (started 1575 B.C,3575 B.P.), which was of Egyptian kings. ("A new king over Egypt, which knew not Joseph." Exodus 1;8). The other question is how long did the enslavement and strife last and when did the Hebrews begin to leave Egypt? We do not know for sure whether they left in one group, as is told in the Bible, or by several emigration waves. If the period of hardship lasted 200 years, Exodus could have happened, for example, at the time of Akhenaten (1362-1345 B.C.), the king who declared the sun-Aten as the chief god over Egypt. It could have taken place during the reign of Tutankhamun (1344-1335 B.C.), the well-known Pharaoh whose tomb was found untouched about a century ago. These were weak kings who, apparently, were not able to organize a campaign outside Egypt in order to force the Israelis back into slavery.

The suggestion or even insistence of the author to connect Exodus with a humid spell in about 1600 B.C.(3600 B.P.) stems from two main, interrelated reasons. In the first, as mentioned already in Chapter 1, the author concluded that there was a climatic change at about this time from the ^{18}O curve in the core sample taken from the bottom of the Sea of Galilee. He then found a good correlation between these results and those reported from the Libyan Desert . Parallel to this study the general paleoclimatic conceptual model for glacial periods was developed, which indicates that during a cold spell rain storms ushered in by heavy dust storms will enter northern Egypt. It then occurred to the author to try and compare the effects of such a climatic change with the story of the "Ten Plagues of Egypt". This comparison is presented as follows.

Floods start running in the wadis or ephemeral river-beds of northern Egypt during the winter seasons. These wadis are mostly dry during normal arid periods; if they run, it is for short periods and not simultaneously. These floods which occur in abundance during humid periods erode and carry silts and soils from the surrounding hills into the clear water of the Nile, causing it to become brownish, or even reddish (the Nile is low and relatively clear during winter). As mentioned, the soils of northern Egypt were called "Deshret", namely "Red". This would spoil the water of the Nile as a source of drinking water and, when heavily laden with silts and soils would kill the fish in the river and in the ponds and lakes adjacent to it. "And all the waters that were in the river turned to blood. And the fish that was in the river died and the river stank; and the Egyptians could not drink of the water of the river; and there was blood throughout the Land of Egypt" (Exodus 7:20-21).

The humidification of the land due to heavy rains would have caused an increase in the bio-world. The habitat of frogs would have expanded beyond the banks of the Nile. "And the river shall bring forth frogs abundantly, which shall go up and come into thine house, and into thy bedchamber" (Exodus 8:3). (On the Sede Boqer Campus, located in the middle of the Negev desert, where the Institute for Desert Research is situated and where the author lives, the introduction of irrigation brought about an explosion in the population of toads).

The humidification of the extremely dry climate of Egypt would also have caused an increase in the number of parasites, such as lice and bacteria causing plagues infesting man and his animals. "And ... the dust of the land ... become lice throughout all the Land of Egypt" (Exodus 8:16) "... and there came a grievous swarm of flies ... the land was corrupted by reason of the swarm of flies" (Exodus 8:24). "The hand of the Lord is upon thy cattle which is in the field, upon the horses ... there shall be a very grievous murrain" (Exodus 9:3). "... and shall be a boil breaking forth with blains upon man, and upon beast ..." (Exodus 9:9).

The storms, coming as rain storms, push heavy clouds of dust ahead of them, darkening the light of day, and are followed by heavy hail, lightning, and a heavy load of static electricity in the air. This would be a very uneasy experience even for a geologist from a more humid region, how much more so for a people who were born and grew up in a land in which the annual precipitation is almost nil. "... and the Lord sent thunder and hail, and the fire ran along upon the ground; and the Lord rained hail upon the Land of Egypt" (Exodus 9:23). "And Moses stretched forth his hand toward heaven; and there was a thick darkness in all the Land of Egypt ..." (Exodus 10:22).

The surprising correlations between what should be the environmental consequences of the new paleoclimatic conceptual model suggested by the author and the description of the Ten Plagues has brought the author to the conclusion that an historical event is, indeed, in the background of the story of Exodus. Although the details of the story were, most probably, different from those of the Bible, yet the general lines portray historical truth.

Describing the story of the travels of the patriarches and Exodus from a paleo-environmental point of view, the story would run as follows: the movement of the desert belt northward beginning at about 4000 B.P. (2000 B.C.) caused the Hebrews to move along the border of Mesopotmia and later southward into Canaan . At about 3750 B.P. (1750 B.C.) the Hebrews joined the Hyksos movement into Egypt in order to escape a drought that caused famine throughout the Fertile Crescent. At about 3600 B.P. (1600 B.C) the climate began to change; while Canaan became more humid due to the movement of the climatic belts southward, it caused Ethiopia to become dryer, which brought a fall in the level of the Nile. This caused famine in Egypt. At the same time it caused dust/rain storms to penetrate into northern Egypt, destroying crops, polluting the Nile, and causing plagues. This gave the Hebrews the opportunity to escape serfdom, as the desert of Sinai, which separated their land of enslavement from the Land of Promise, became less hostile as the more frequent rains over this land held the promise of an increased water supply and better grazing.

After the story of the "Ten Plagues" comes the story of the crossing of the Sea of Reeds (or Red Sea as it is called today), followed by the tale of the wanderings in the desert of Sinai. Although there is no intention of drawing a map of the route the tribes of Israel took through the desert, still the general characteristic of the area seem to fit well with the stories of their adventures along this route

In this chapter the details of the biblical narrative of these event will be examined and compared with knowledge accumulated on the Sinai and its water resources.

First, it should be clear that the term "sea" in the Bible refers to a body of standing (as opposed to 'flowing') water. Thus lakes and even large water tanks, like that of Solomon's Temple in Jerusalem, were called seas. Second, the sea which the tribes of Israel crossed on their way from Egypt to Sinai is called Yam Suf, the Sea of "Reeds". This means that reeds grew along its shores, which must have been fed by a source of fresh or brackish, but not sea water. The author believes that this "Sea of Reeds" consisted of the huge swamps and lakes (locally called "Sabkhas"), which at that time spread along the rift valley between the Red Sea and the Mediterranean Sea, through which the Suez Canal presently passes. When the tribes reach the other side of Sinai, they come again to Yam Suf. This time it is the other branch of the rift, namely the Arava valley (Figs. 9.1, 2.4). These sabkhas were, and still are, formed by groundwater emerging from deeply buried aquifers to reach the surface of the land. Such deep aquifers, containing brackish water under artesian pressure, were discovered by oil and water exploration wells drilled in this region. It is thus fairly safe to assume that seepages of this water emerged in those days (as they emerge today at the southern edge of this area, at the springs of Uyun Musa), along the central and northern part of the fault, forming swamps surrounded by reeds. Moreover, the sands covering the area east of the rift valley also contain a shallow water table of brackish water, which could have risen during a spell of humid years to form marshes, and even lakes, surrounded by reeds. Such lakes are found today during a very humid winter, in the valleys between the sand dunes east of the Suez Canal. Thus, it seems very probable that a more humid period will create an environment of marshes and lakes in the region separating Africa from Asia.

However, it should not be forgotten that humidity in such a region is a very relative term compared to other regions. Even during a humid spell this region was still arid; precipitation, even if double that of today, would not exceed 100 mm. In such an arid environment, sabkhas or swamps are characterized by the formation of a crust of evaporates such as calcium carbonates, gypsum, or halite (table salt).

According to the chemical analysis of the water found in the deep aquifers and springs, the ancient sabkhas along the sea shore were probably rich in calcium sulfate (gypsum). Such deposits are found today in this region. In the areas where seepage of brackish water from great depths occurs today, as in the vicinity of the springs of Uyun Musa and Saidna Musa emerging along the faults bordering the Suez rift on its eastern side, sabkhas with an evaporitic crust of gypsum are common. Another chemical component of this water is magnesium sulfate which gives it a bitter taste.

With this hydrogeological and hydrochemical background, it is now suggested that we try to reconstruct the flight of the tribes through this area. Thus, we might now consider a large group of people coming from the fertile Land of Egypt, trying to escape the Egyptian army. The season is spring which is typified, even today, by rainstorms following hot, dry spells or "Khamsins." which are dry winds heavily laden with dust, blowing from the east

We are told that the tribes leaving Egypt preferred not to take the northern, shorter route along the coastline, as this was the "Road of the Kings" or the "Land of the Philistines" (although the "true" Philistines did not arrive until 1200 B.C. (3200 B.P. during the reign of the 20th Dynasty); the reason for not taking this route being, as archeological evidence shows, that this road was lined and patrolled by Egyptian army garrisons. It was called by the Egyptians "The Way of Horus" (Fig. 8.2).

The tribes took a more southern route, although this route ran along the shore it was shorter and strewn with many oases between the sand dunes. These oases are supplied with good water (coming from rainwater, which, although scanty, infiltrates rapidly through the porous sand and accumulates on the impervious underlying clays of Neogene age). The southern road brings them into an area infested with swamps, salt marshes, sabkhas, and thickets of reeds. Reaching this area, they hear the report that the chariots of Pharaoh are following them, to march them back into slavery. They encamp along the western border of these stretches of swamps and marshes. At that time a fierce dust storm, typical for this time of the year (but probably even more fierce than in our time, due to the different climate) blows from the east, reducing the visibility to nil and stoping the Egyptian army from overtaking them as "the pillar of cloud went from before their face and stood behind them" (Exodus 14:19). The same hot, dry storm dries up the gypsum and salt crust of the soil surrounding the swamp, and the sea goes back "by a strong east wind all that night and made the sea dry land and the waters were divided" (Exodus 14:21), most probably by a hardened crust that formed by a layer of gypsum. This might have given the Israelites an opportunity to escape the Egyptian army by going westward into the maze of marshes. When the Egyptian army moved after them it could not envelope them from the flanks, as any cavalry or chariot army would have liked to do to a swarm of pedestrian people, because of the marshes which "were a wall unto them on their right hand and on their left" (Exodus 14:22). In this context, a wall would mean a protection on their flanks. The chariots could only move behind them and try to overtake them on the dried-up tracks. "And the Egyptians pursued and went in after them to the midst of the sea, even all Pharaoh's horses, his chariots, and his horsemen" (Exodus 14:23). At this point two phenomena might have taken place. First, the crust which would bear the weight of man could not carry that of the chariots, the wheels of which started to sink into the crust (an experience the author has had two or three times in sabkhas of this type, until he learnt the lesson that where a geologist can walk, a jeep may sink up to its shaft). This sinking delayed the Egyptian army in the midst of the sabkha. "It took off their chariot wheels that they drave them heavily" (Exodus 14:25). At that time the khamsin may have "broken", namely, the dry, hot eastern wind was the precursor of a low pressure rainstorm coming from the sea and flooding the area. Thus "the sea returned to its strength when the morning appeared and the Egyptians fled against it (Exodus 14:27).

This scene could have happened along any part of the valley between the present cities of Suez and Port Said, as all this area was most probably infested by such swamps. A rise in the groundwater table in the water-bearing layers adjacent to the Suez Rift Valley is also a possibility. This caused the area to become more swampy, with many springs of brackish water so that reeds grew around and in the midst of the swamps. Furthermore, the regime of heavy dust- and rainstorms following one in the wake of the other was much more intensive than at present. If one adds to this the relatively high gypsum content of the water emerging from the aquifer and brought by the floods, and the drying effect of an eastern dust-laden wind ushering in such a rainstorm, the stage is set for a scenario such as that described in the Book of Exodus. As previously mentioned, one should remember that a regression of the sea due to water caught in the glaciers which spread in the polar regions during that period could also have changed the geography and environment of the area between Egypt and Sinai, as well as all along the shores of Egypt. A regression of the sea in the order of magnitude of a few meters would have changed all the travel alternatives between Egypt and Sinai.

This is another reason why the author is reluctant to draw a map indicating the routes the Israelites wondered in Sinai. The first reason has already been stated, namely, that the author does not suggest taking the stories of the Bible as scientific reports, but they are only an echo of the traumatic experience the people went through. Moreover, it is claimed that the fiercer the experience, the more it will portray the environmental background. (The reader who is interested in the routes of Exodus is advised to consult [1, 6]).

Fig. 8.2. The Egyptian line of forts along the "way of Horus" from Amon's temple in Carnak, from the time of Seti I (1291-1309 B.C.). Each fortress is provided with a pool of water

Assuming a change of climate did take place, another question which comes to mind is how it influenced the Desert of Sinai? From the Bible, the echo of a journey through a formidable desert, the suffering from lack of water, is heard. How do these echoes agree with the assumption of a change in climate? The answer to this lies in understanding the nature of such a change and thus its impact on the desert. The change which is assumed would not have changed the desert into a humid place, but would have produced a special type of climate decided by a general southward movement of the arid belt. Yet, during summer this belt has its seasonal movement northward. Thus, this region still maintains its arid character of long, hot summers which might have been interrupted by a few rainstorms. During autumn, winter, and spring, frequent spells of rainstorms following heavy dust storms would occur. This is due to the seasonal movement of the arid belt southward. From the ecological, environmental point of view, such a climate might affect the quantity and frequency of the floods between autumn and spring, increase the amount of water flowing from the springs, and increase the vegetation cover, especially in river-beds and in the mountain regions. Thus, more grazing and firewood can be found in the desert. It would not, however, make the desert a fertile land or more pleasant, especially during the summer months. The author is of the opinion that a picture of this type of desert is drawn in the story of the wanderings of the Tribes of Israel in the deserts of Sinai and the Negev.

9 That Great and Terrible Wilderness (Deut. 1:19)

> So Moses brought Israel from the Sea of Reeds and they went out into the wilderness.... and found no water. And when they came to Marah, they could not drink the waters, for they were bitter; therefore the name of the place was called Marah (Exodus 15:22,23)

The attention of the reader is drawn to that fact that the Biblical story mentions the bitterness of the water, and not its saltiness. A bitter taste in drinking water in these arid regions is, in most cases, a function of the relative abundance of sulfate salts, especially magnesium sulfate ($MgSO_4$) (Epsom or English salt) which, even as a small constituent of the salts, will give the water a bitter taste.

The large springs of Uyun Musa ("Springs of Moses" in Arabic), which can supply drinking water for a multitude of people, are located on the northeastern edge of the Red Sea, not far from the northeastern bank of the Gulf of Suez (Fig. 8.1). The water in these springs is rich in sulfate (500 ppm) mainly calcium sulfate, but also contains magnesium sulfate which gives a bitter taste.

Although the author does not consider the biblical story a scientific report, he still suggests that attention should be paid to the fact that the story emphasizes the bitterness of the water. This emphasis distinguishes the story as one of the first reported "hydrochemical" observations in human history. If one accepts the identification by travellers and geographers of Uyun Musa with Marah, and continues to investigate the secret behind the emergence of these large springs on the border of the desert, one will arrive at a very interesting hydrogeological story which began about two and a half million years ago and still plays an important role in the environment of these deserts.

The best way to understand the impact of the hydrology on the environment is to view the region from space, using satellite photographs or images produced by NASA. These give a good picture of the aridity of the deserts of Sinai and the Negev (Figs. 2.4, 8.1). One can see the dark colored exposures of the rocks with no green cover from forest or woods in the mountain region of southern and central Sinai and the Negev. The plains are bright colored with very sparse vegetation. The satellite images render a more detailed picture. In the image the reflection from the earth's surface of different wavelengths of light can be read as different colors, on a false-color image picture. The red color, in most cases, means the reflection of light from an area with a vegetation cover. A more detailed study of the Landsat images of Sinai and the Negev deserts show such specks of vegetation along the fault lines bordering the rift valleys (Figs. 2.4, 8.1). In deserts, a concentration of vegetation denotes places where groundwater emerges.

In the previous chapter, the emergence of groundwater along the rift valleys of the Suez and Arava was mentioned. The following is a description and explanation of these phenomena in more detail and thus to understand better the impact of the availability or lack of water resources on the fate and faith of the tribes travelling through these areas.

Looking at these satellite pictures and images of Sinai, one's attention is drawn to the triangular shape of the Sinai pininsula (Egypt) and the Negev (Israel), which together form a transition zone between Asia and Africa. One can observe that the two regions, divided by the political border between Egypt and Israel, cannot be distinguished from each other from a vantage point in space. In fact, they form one unit, from both a geological and a geographical point of view. Also, both areas have features in common with the regions bordering them to the north and the west: the Sinai peninsula is similar to the eastern and western deserts of Egypt, and the Negev is similar to southern Israel, Jordan, and northern Arabia.

This resemblance signifies the special character of the Sinai peninsula and the Negev. These regions existed as one geological unit during most of their geological history and were divided from Asia on the east and Africa on the west only comparatively recently. The division was effected by the large Syrian-African rift system, with the Gulf of Suez rift on its western edge and the Gulf of Elat and the Dead Sea rift on the eastern edge. This separation of Sinai and the Negev from Africa and Arabia occurred at the dawn of the Quaternary era, some two and a half million years ago, and was the culmination of plate movements which affected this area during the Neogene when a shallow sea penetrated into the down-folded and faulted, elongated basins.

The deep troughs, or rift valleys, formed by the shearing and the down-faulting movements, continuing through the lower and Middle Quaternary, gave the rift valleys their present morphological features and caused the diversion of the drainage patterns from the Mediterranean Sea to the rifts. This rifting also profoundly affected the subsurface flow of groundwater. The regional faults along the rift valleys' borders fractured the rocks along the faults and vertically displaced impermeable layers, thus opening flow routes between the deep subsurface and the surface. Water still uses these routes to flow from the depths to the surface to form springs.

Due to their in-between character, the Sinai and the Negev maintain geologic, structural, and stratigraphical affinities with the African and Arabian plates, and can be regarded as a miniature plate which exhibits, on a small scale, many features of the larger plates. On the southern edge of the Negev-Sinai peninsula, one finds a triangle of the basement shield, built mainly of crystalline rocks of Pre-Cambrian age (Figs. 2.2, 2.4,). This part of the Sinai is mountainous. Mount Sinai is traditionally located in this region. Northward, an area built of sandstones overlies the crystalline basement. These sandstones were deposited after the lower Paleozoic or mid-Mesozoic on a wide continental stable zone, seldom invaded by the sea. When sea intrusion did occur, thin shallow marine deposits were left behind.

The sandstone rocks are highly permeable and form good aquifers (water-bearing strata) when water infiltrates them. The belt of sandstone layers is bordered to the north by a steep escarpment formed by limestones of Middle Cretaceous age. These are overlain by chalks of Upper Cretaceous age and limestones and chalks of Tertiary age. The plains of central Sinai are rolling hills with occasional mountains protruding from the plain. The hills are built from nearly horizontal layers of chalks and marls of Upper Cretaceous to Tertiary age, while the mountains are anticlinal structures of folded limestones of Middle Cretaceous age. In some of these anticlines, the cores have been eroded and sandstones and limestones of Lower Cretaceous to Jurassic age are exposed . The chalks and marls are impermeable to water, and thus springs are scarce in the plains of central Sinai. The valleys between the hills are filled mostly by desert loess (an airborne silt deposited during Upper Quaternary times) and some gravel and sand layers. The most northern belt of the Sinai peninsula and its extension into the Negev is formed from a plain covered by sand dunes (Figs. 1.1, 8.1, 9.1).

The scanty rains (50-100 mm per year) falling on these dunes infiltrate quickly into the sands and thus do not evaporate. This water accumulates on the impermeable layers underlying the sand dunes and thus forms a shallow groundwater table with a gentle gradient toward the sea. Near the coast, the groundwater table is very close to the surface of the land, and where it outcrops, or where the inhabitants dig shallow wells to reach this water, oases are found. These oases determined the routes taken by people crossing the deserts. The ancient routes connecting Egypt to the Fertile Crescent of the Middle East were along the coast of the Mediterranean, where there are strings of oases fed by the shallow perched groundwater table. El-Arish is the main oasis on this desert crossing. The groundwater here is also fed by the occasional flooding of Wadi El-Arish, which drains most of central Sinai. The water spreads over the flood plain of the wadi, percolates to the subsurface gravel beds, and from there to the sandstones and sands of Quaternary age.

A well-guided traveller in ancient times could depend on a reliable supply of water for his party and his pack animals over most of the northern route. He would have to carry only enough water to help him cross some of the longer stretches between the oases. However, crossing the Sinai desert becomes more difficult for even small groups of travellers, not to mention large groups, if they must choose a more southern course. Water resources are more scarce along this route than in the north. Only two clusters of large springs, sufficient to supply a multitude of people, are to be found. One is the Uyun Musa springs near the Gulf of Suez (Figs. 8.1, 9.2) and the other is the Ein Qudeirat, Qadis and Quseima spring cluster (apparently the Biblical Kadesh Barnea) near the border of Israel (Figs. 9.1, 9.3).

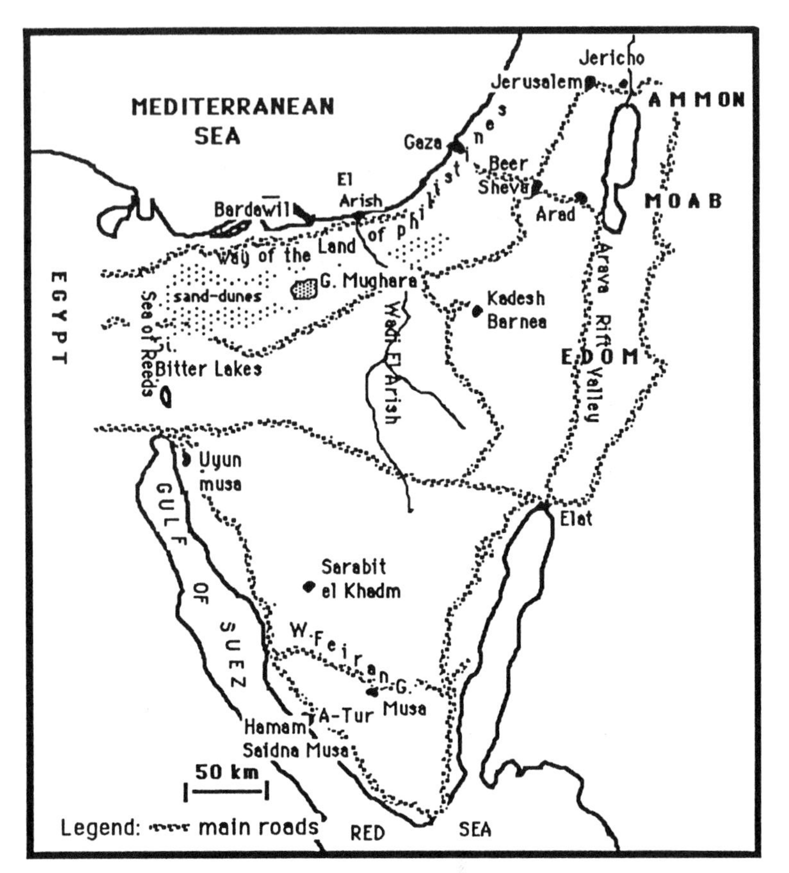

Fig. 9.1. Routes through Sinai

The traveller from Egypt to the Fertile Crescent, who crosses Sinai through its central areas, will reach the rift valley of Suez on his way eastward. However, he will have to take a southerly path if he wants to provide himself with water before entering the Sinai desert. If he travels a few tens of kilometers southward, he will come to a group of springs with an abundant quantity of water. Near the springs is an extensive marshy area, partially covered by thick clusters of reeds, tamarisk trees and date palms. The name of these springs is Uyun Musa which means "Springs of Moses". There is something peculiar about these springs. A few of them appear at the top of small mounds built of calcareous and gypsiferous tufa (Fig. 9.2). The mounds were created by the springs themselves. The spring water contains calcium, magnesium carbonate, and sulfate which were precipitated when the water emerged and flowed on the surface. Since the mounds are a few meters high, the springs must have existed for many thousands, if not tens of thousands, of years. The large quantity of water which continues to flow at present, forming an extensive marsh, indicates that a rich aquifer has supplied these springs throughout the years without being exhausted.

What is the secret behind this rich flow of water at the edge of the desert? A short survey will show that these springs are not fed from local rains or floods. The layers out of which the water flows and on which the spring tufa mounds are built are limestones and marls of Neogene age. The extension of these rocks is limited, and a small quantity of rain falls on its outcrops, if it rains at all. Therefore, hydrogeologists investigating groundwater resources of this desert believed that the water's source might be a deep aquifer which stores a huge quantity of water in its pores. The water would be pushed up to the surface by artesian pressure. A change in this pressure may be effected by a change of climate, namely, by more rains falling on the outcrops of the aquifers in southern Sinai and the water infiltrating the subsurface, raising the water table (Fig. 9.4).

Fig. 9.2. One of the springs of Uyun Musa. The outflow is from the top of a hill built of calcareous-gypsiferous tufa. (Photo by E. Mazor, Weizmann Inst., Rehovot)

Fig. 9.3. Air photo of the oasis of Ein Qudeirat (Kadesh Barnea). The spring outflows (at the upper left part of the picture) from the limestones of Eocene age which builds the surrounding hills. The ancient 'tel' can be seen at the upper part (Photo by Scientific Survey, Sinai)

Such artesian aquifers are found under many of the deserts of the world, such as the Sahara [1, 2 ,3]. In many instances their outlets are along fault lines. This water can be defined as "fossil-" or "paleowater" since it is many thousands of years old. The age was determined by the ^{14}C content of the carbonate dissolved in the water. (For an explanation of the role of isotopes in hydrological research see Appendix II). In most cases the age is in agreement with the time-span calculated on the basis of hydrologic flow models. Such a time period would have been necessary for water which had infiltrated the subsurface in the outcrop areas to flow to the desert oases, located hundreds or even thousands of kilometers away. Yet pressure waves will travel faster. These will affect the flow of the springs thousands of years before the recharged water will reach the outlet. One can imagine the difference between the velocity of the flow of water and the velocity of the pressure wave which affects the flow of the springs as the difference between a sound wave and the velocity of the wind.

It is no wonder that Moses led the tribes toward these springs after entering Sinai. He knew from his previous wanderings in the desert that there was an ample supply of water. Here, however, the people were bitterly disappointed as the water tasted bitter, especially after the fresh Nile water to which they had been accustomed to all their lives. We do not know how Moses managed to sweeten the water. Based on personal experience, the author knows that, as the thirst grows, one loses ones sensitivity to the bitterness of the water, and as one continues to drink the water for a long while, one loses even the sensitivity to the maleffects of the magnesium sulfate in the water. (For an explanation of the sulfate anomaly in the water, refer to Appendix III).

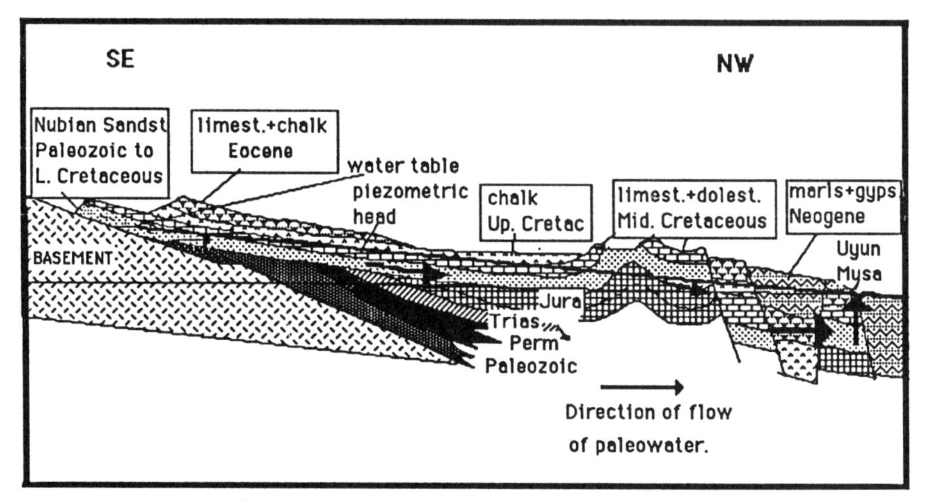

Fig. 9.4. Hydrogeological cross section through the Sinai Peninsula

After becoming acquainted with the landscape of the Desert of Sinai, and its geological and hydrological features, it is suggested to investigate the problems facing any leader or commander who must plan the crossing of this desert by a multitude of people. The first problem, undoubtedly, would be the supply of water and food for the people and their animals.

Taking into account the possibility that the climate was more humid, with a precipitation perhaps twice today's, this would raise the quantity of rain from 80 to 160 mm. One can assume that during winter, spring, and early summer, most of the springs would flow abundantly and enough forage could be found on the desert plains. Problems of supply would have to be faced, either when a large concentration of people stay in one place or towards the end of the summer, when small springs may fail and the desert grasses are depleted. Problems of water supply may occur on the long stretches between one group of springs and another, or when a year or two of drought may deplete the flow from the springs with a small storage in their subsurface. In such cases, the leader would have to disperse his people to try and find springs and fodder in the higher mountains where more precipitation, and even snow in winter, promised more water and food. Another problem to be overcome is that described in Marah, namely, the bitter taste of water caused by the sulfates. However, such springs are encountered only along the rift valley, namely, at the beginning of the wanderings and might also have been encountered toward the end before crossing into Moab.

Thus, analyzing Sinai from the hydrogeological and environmental point of view, one would say that only few places exist suitable for the concentration of many people (Fig. 9.1). The first place is around Uyun Musa, where the brackish springs emerge. Another is between the sand dunes northeast of Uyun Musa. A third is in the vicinity of the present small town of A-Tur where a warm brackish spring emerges, the name of which is Hamam Saidna Musa (The Bath of our Lord Moses). Here, many palm groves flourish because of the shallow freshwater table that exists in the vicinity of these springs. Also, in river-beds between Uyun Musa and Hamam Saidna Musa, clusters of small springs are found, supplying sufficient water for small groups of people, but not for a multitude over a long period.

An area which might have been rich in springs is the erosion cirque at Gebel Muaghara, where rocks of Nubian sandstone are exposed and where a shallow water table also exists. Another place would be in the south , namely in the region built of crystalline rocks, such as granites, gabbros, and porphyry (Fig. 2.4). There are many small springs and oases in this region, the biggest of which is Wadi Feiran where many groves of palm trees grow.

As mentioned, another group of perennial springs and a large area with shallow groundwater is found at Ein Qudeirat and the plain of the small town of El-Quseima (the site of Kadesh Barnea). This place is mentioned as the main base of the tribes of Israel and their gathering place before their first attempt to invade Canaan via the Negev .

According to the Bible, after leaving the springs of Marah and camping at several watering places, the tribes of Israel reached Mount Sinai, where they had a big gathering. During this time, Moses went up the Mountain for the Tablets, the receipt of which was connected with thunder, lightning, fire, and the loud call of a trumpet.

South of the spring of Uyun Musa, down to A-Tur (Fig. 9.1), the area is densely faulted; tilted blocks of sedimentary rocks conceal rich oil deposits, some of which seep up to the surface. It is quite possible that during the time the tribes were in this region an earthquake could have caused the emergence of oil and gas which caught fire and blazed. For centuries, such phenomena were known to occur in western Persia and were connected with fire worship. Another experience might have been a severe electrical storm of lightning and thunder. Such a spectacle descending upon a high mountain in the middle of the desert is very impressive. Such a storm, as witnessed by the author, may cause the atmosphere to be loaded by a static electrical charge which may be discharged through the human body giving him a rather unpleasant shock. Thus, a combination of a few dramatic natural phenomena occurring while the tribes were in the area along the Gulf of Suez could explain the stories narrated in the Books of Genesis and Exodus.

The Egyptian center of worship to the cow-goddess, Hathor, in the turquoise mines of Sarabit-el-Khadm where some of the tribes would have tried to re-adopt rites they remembered from the country they had left so recently, provides an understandable background for the Biblical story of the golden calf (Fig. 9.1). Although a humid phase is supposed to have occurred during the final stage of the stay of the Tribes of Israel in Egypt, the stories of the wanderers' suffering from thirst in the desert are not surprising. As already explained, the average annual precipitation over most of Sinai is less than 100 mm and, in many areas, is even less than 50 mm. Thus, even a humid spell which might have doubled this quantity would still have produced only fairly low precipitation. Moreover, a dry spell may have occurred after the tribes penetrated the desert. The tribes may thus have arrived in the area expecting to find flowing springs, to be surprised by dry rock. In such cases, the most advisable thing would have been to dig into the rock to locate the fractures which supplied the water to the springs. These fractures might have contained water which would then fill a shaft excavated into the rock [4].

Fig. 9.5. Air photo of Gebel Musa, Central Sinai (Photo by Scientific Survey, Sinai)

This procedure, which most probably was also known to the ancient inhabitants of this area, is now applied by the local bedouin tribes occupying the crystalline province of Sinai (Figs. 2.4, 9.5). To locate a dried out spring one must first of all find the travertine layer deposited by the disappearing spring (Fig. 9.6). This is a crust formed by the evaporation of water in which lime has dissolved. When such a crust is found on a fracture in the crystalline rock, the fracture is followed to the point where the hard granite or porphyry rock is traversed by soft gabbroid or diabase rock, usually decomposed into clay (Figs. 9.7, 9.8). A shaft is then dug into the decomposed rock while care is taken to follow the fracture along the wall of the shaft. At a certain depth the water will start to seep from the fracture into the shaft. In some places, the removal of the lime travertine from the fracture is needed to enhance the flow. Thus, the story of the striking of the rock to obtain water is not altogether alien to the situation in southern Sinai. It would undoubtedly have looked miraculous to people coming from a land where the only source of water was the big river and the canals dug along its banks.

Fig. 9.6. Layers of travertines in fractured granites (Photo by A. Ecker, Geological Survey, Israel)

Fig. 9.7. Monastery located at the outlet of a small spring flowing from a fracture, bordered by a negative dike (Photo by the author)

The spring mentioned quite frequently in the Bible as a camp of the tribes of Israel was Kadesh Barnea. From here the mission headed by Joshua and Caleb went to spy the promised land and from here the first attempt to storm the land was made. Kadesh Barnea is an ideal concentration place for people in the middle of the desert. A large spring issuing from the limestones of Eocene rocks emerges at the head of a valley. This spring is fed by the rains and floods flowing over the rocks building the anticline of the Ramon, which is the highest mountain of the Negev (reaches 1000 m above MSL) and thus receives more precipitation. The water infiltrates into solution channels in the limestone rocks of Middle Eocene age until it reaches impermeable chalk layers of Lower Eocene age and marls of Paleocene age, on which a regional perched water table is formed (Fig. 9.3). In addition to the main spring of Kadesh Barnea, the special geological conditions cause many small springs to seep from the rocks in this area. One of them still has the name Ein Qadis. The outlet of the narrow valley of Kadesh Barnea is into a broad valley where the small town of Al-Quseime lies. Here, many small brackish springs seep from the ground allowing many shrubs, palms and grasses to grow. Today the spring of Ein Qudeirat flows at about 40 m^3 per hour. This quantity may suffice to supply drinking water for a few tens of thousands of people providing they water their stock from the brackish springs in the lower valley. Taking into account that a more humid period would have caused an even larger flow, one can see the sense in the decision to make Kadesh Barnea the pivot point for the wandering tribes.

Fig. 9.8. Pumping a well which was excavated on a fracture, granitic province of Sinai (Photo by A. Ecker, Geological Survey, Israel)

It is more difficult to understand the story about the need to strike the rock to get water at Kadesh. Was the story from another place transferred to the more permanent location of the tribes or did the spring of Kadesh also fail, thus necessitating excavation into the rock? According to records known to date, this spring did not dry up even after a series of a few dry years, though its flow did diminish. The possible explanations thus are either a more conservative one, namely that the last stages of the stay at Kadesh Barnea a dry period restarted, which caused the springs to dry up and which necessitated the digging of wells. The same dry period forced the Israelis to try and break out from the desert into the more humid lands. The less conservative approach will suggest that this story was attached to this place by the writers and editors of the Bible in some later period. As already mentioned several times, the purpose of the present study is not to explain biblical and mythological phenomena but to offer a better understanding of the environmental background which may have caused them to happen and be remembered in one way or another. Thus, within the boundaries of this assignment, it can be claimed that, while with the better knowledge of the bedouins' methods of excavating for water in the crystalline rocks, the story of Moses' striking of the rock can be understood, it does not fit so nicely with the hydrogeological situation at Kadesh Barnea.

The base at Kadesh Barnea saw the first attempt of the Israelis to enter Canaan by force and their first defeat. They were beaten back by the King of Arad. This story raises some questions. The archeological excavations at the site which was definitely identified as the ancient town of Arad, showed that this town was flourishing as a strong fortified city during the Lower part of the Early Bronze Age. It was abandoned at about 4600 B.P. (2600 B.C), and was not rebuilt until the Iron Age. Thus during the Upper Bronze Age, the assumed period when the Israelis arrived, the town was abandoned [5].

In the opinion of the author of the present work, there exists no real controversy between the story of the Bible and the archeological findings. In general terms, it can be said that the history of this site as revealed by the excavations fits well with the paleoclimatic story explained in the previous chapters. The flourishing of the city during the time of the Early Bronze Age coincides with the wet climate period which continued from the Chalcolithic period, while its decay and abandonment at about 4600 B.P. is in agreement with the desiccation of the climate, which can be seen on the ^{18}O curve of the core samples taken from the bottom of the Sea of Galilee.

While the city of Arad was still occupied, its people excavated a deep shaft in the lower stretches of the city reaching into a shallow groundwater table. Although the archeologist Ruth Amiran (personal communication), who excavated the site, maintains that the well is from the Iron Age, the present author disagrees with this conclusion, as he maintains that the archeological considerations did not take into account the fact that a humid spell could have caused the water table to rise and thus in the early stage a rather shallow well could have sufficed for the town's water supply. During the Iron Age, when methods of digging wells improved, the well was enlarged in diameter and deepened. Thus only remains of the last period were left at the bottom of the well.

This well ensured a supply of water in times of siege. As the area replenishing this water table is rather limited and does not greatly exceed the area of the town, it seems probable that a series of dry years and the dense building up of the replenishment zone brought about the drying up of this project and the desertion of the city by most of its people. Spells of wet years during the Upper Bronze period caused the water table to reappear and, as the site of the well was known to the semi-nomads who dwelt in the surrounding area due to tradition passed from father to son, they would have again gathered around it. Thus, it can be understood that the site of Arad and its well, like other sites in the vicinity where other shallow wells exist, were settled by people headed by a chief, called the King of Arad, This chieftain controlled the passes from the Sinai and Negev deserts to the mountain of Hebron and was able to push back the people who tried to invade this fertile land.

The role of similar deep wells in the Canaanite cities, excavated to ensure their water supply, will be discussed in more detail in the following chapter.

At some stage, the Israelis decided to avoid a direct assault across the desert border, where too strong an opposition was encountered, and to circumnavigate it. They tried to march through the Land of Edom, most probably the present Central Negev, but were denied free passage. They then pressed southward toward the Gulf of Elat, crossing somewhere north of the gulf, to reach the more humid heights of the mountains of Moab and Ammon where they avoided war by abstaining from harassing the local inhabitants whom they considered kin. They moved northward, defeated the Amorites inhabiting the Gilead and Bashan heights and were ready to cross the Jordan. It is interesting to note that after they had crossed the rift valley and gone up the mountains of Trans-Jordan, no more complaints of thirst were reported. During this part of their journey the tribes experienced an event connected with the excavation of a water well for which occasion a special hymn was composed.

> "Spring O Well, sing ye unto it. The princes digged the well, the nobles of the people digged it, by the direction of the lawgiver, with their stave." (Numbers 21:17,18)

10 Wars and Rivers, Walls and Wells

> And as they that bare the ark were come unto Jordan, and the feet of the priests that bare the ark were dipped in the brim of the water, for Jordan overfloweth all his banks all the time of harvest. That the waters which came down from above stood and rose up one heap.... (Joshua 3: 15,16)

> The kings came and fought, then fought the kings of Canaan in Taanach by the waters of Megiddo; they took no gain of money. They fought from heaven, the stars in their courses fought against Sisera. The river of Kishon swept them away, that ancient river, the river Kishon. (Judges 5:19-21)

The first move of the people of Israel into the Promised Land is reported to have been accompanied by an anomalous hydrological event. The Jordan stopped flowing and enabled the tribes to cross its bed without wetting their feet. Some scholars have suggested that this might have happened when a landslide of the soft marls of which the banks of the river are built caused damming of the flow, thus enabling the tribes to cross. Taking into account the lithological character of the rocks forming the banks of the river, and that there are known historical records of the occurrence of such events, this explanation seems feasible. The last historical record is from 1927, when a severe earthquake caused much destruction all over the country. A survey along the Jordan's banks, from the Sea of Galilee to the Dead Sea, shows that the breadth of the river is many times that of the height of the walls. One of the few places where such a damming can occur is near the river pass at Tel Damie, presumably the city of Adam; the very place mentioned in the Bible as that at which the water stopped and stood as a wall. The discussion of the drying-up of the Jordan gives the opportunity to introduce the reader to the hydrology of this famous, though small, river (Figs. 6.2,10.1).

The Jordan in its present form is a relatively young river, not more than about 10,000 years old. Before this period, all of what is now called the Jordan Valley

Fig. 10.1. The Jordan south of the Sea of Galilee. The hills overlooking the river are built of the soft Lisan Formation marls, of Pleistocene age (Photo courtesy Irit Zaharoni, Bamahne Journal, IDF)

was covered by a long lake extending from north of the present Sea of Galilee to south of the Dead Sea. The level of this lake reached an absolute altitude of about 200 meters below sea level. It was formed as a result of the humid climate which prevailed in this region during the Last Glacial period. At about 14,000 years B.P., the level of the lake began to recede slowly and at about 10,000 years B.P. it reached its present form of two main lakes: the Sea of Galilee and the Dead Sea. A very shallow lake surrounded by marshes existed until recently in the Huleh Valley. The regression exposed the layers which had been deposited in the lake since its formation about 100,000 years ago. The layers are soft marls, rich in diatoms and freshwater shells in the north and gypsum in the south. The layers are built of fine varves formed by the seasonal variation of deposition in the lake. During the winter, streams carrying floodwater deposited fine sand, while during the summer, when evaporation was high, calcium carbonate and gypsum were deposited. The ratio of gypsum and salt becomes higher the further south one goes. The reason for this is that even at that time, the southern stretches of the country were more arid and evaporation was high. This caused the lake to become brackish to salty in its southern part, although it did not reach the salinity of the present Dead Sea. The softness of the layers exposed caused them to be easily eroded forming whitish colored vertical cliffs. These form the peninsula or "tongue" of the Dead Sea, called "El Lisan" (Fig. 4.1).

When the lake was slowly desiccated, the Jordan cut its way from the Sea of Galilee southward. Due to the softness of the layers the river encountered on its way from the Sea of Galilee (at about -200 meters below sea level) to the Dead Sea (at about -400 m below MSL), and to the rather long distance, the river formed many meanders. The lower banks of the river, along and between the meanders, were always covered by dense vegetation, giving shelter to wildlife and fugitives from society.

There are several sources for the water of the Jordan river. The two main springs, the Dan (Tel Dan-ancient Laish-is situated near this spring) and the Banias (named after the god Pan), emerge from Mount Hermon which is built of limestones of Jurassic age. Deep solution channels, formed by the water containing CO_2 dissolving the calcium carbonate rocks, characterize the landscape of the Hermon. Due to this dissolution process, the limestones are highly permeable and the water from the rain falling on the mountain and from the melting snow which covers the higher stretches of the mountain each winter quickly infiltrate the subsurface to enrich the aquifer which gives rise to these springs (as well as to many other springs such as those irrigating the oasis of Damascus). The average annual total amount of precipitation may reach 1200 mm. The Dan and Banias springs, together with some smaller springs flowing from the flanks of the Hermon, contribute most of the water to the upper Jordan. Due to high permeability and the high rate of precipitation, the water flow of these two major springs is fairly regular. The difference between summer and winter is regulated by the large underground storage of Mount Hermon. A long spell of dry years and a low snowfall on the mountain may cause a decrease in the total quantity of water in the springs, which will cause a reduction in the flow of the Jordan to the Sea of Galilee. These springs flow into a wide valley called the Huleh Valley, the outlet of which was during the Pleistocene by the lava flows from the volcanoes that formed the Golan Heights. The Jordan cuts a narrow gorge through the basalt rock, enabling it to continue its flow to the Sea of Galilee. The damming by the basalt flow caused the formation of a shallow lake and marsh in the Huleh Valley, called "Mei Merom" (Water of the Height) in the Bible. This lake was drained in the late 1950s and its bottom is now agricultural fields.

Many small springs join the Jordan on its way to the Sea of Galilee and floods from the Galilee and Golan Heights join its flow in winter. The Jordan enters the Sea of Galilee forming a small delta where dense vegetation flourishes on the many water streams. The abundance of streams and vegetation formed an ideal place for wildlife and hence the name "Bethsaida" (The Place of the Hunt).

The Sea of Galilee is a regulating basin for the flow from the springs of the north, as well as for the floods and springs from the Golan Heights and the Sea of Galilee. The most famous springs are those of Tabgha near Kefar Nahum (Capernaum) and the thermal mineral springs of Tiberias.

The Jordan leaves the Sea of Galilee at its southern end where it is joined by the Yarmuk river which drains the heights of Golan and Gilead. This river carries mainly floodwaters. Its flow is high during the winter but low during the summer. The hot springs of El Hammah (the Biblical "Hammat-Gader") regulate its flow during the dry period.

From there southward, the Jordan meanders through the marly Lisan layers, receiving spring floodwater from the west and east ephemeral wadis, such as the Jabbok, which flow during the winter and dry up in summer. Thus the flow of the Jordan river, south of the Sea of Galilee is very much affected by the floods of the Yarmuk and the other ephemeral springs. In winter it is a gushing river, very dangerous and difficult to cross, while during the summer, and especially in the autumn, the flow diminishes extensively and the river can be forded in many places. This is the case, particularly after a series of dry years, which causes a drastic reduction of the flow from the Sea of Galilee. During such periods the water level of both the Dead Sea and the Sea of Galilee recedes.

The story of the Bible claims that the fording of the Jordan by the tribes was during the "season of the harvest of the wheat" which is about May. This is the period when the floods of winter have receded but the river has not yet reached its lowest level.

At this time of the year the fording of the river is possible in a few places, the water being about waist height. Thus the claim for a dry passage can be explained by the collapse of a large block of marl into the river, as the result of an earthquake. Another possibility, assuming that the report is fairly accurate, is that the fording took place after a period of dry years which affected the springs of the Hermon and the filling of the Sea of Galilee by floodwaters.

A series of dry years could also be suggested as a triggering factor that made the Israelites decide that they had had enough of roaming the wilderness of Sinai and the time had come to enter a better watered, more fertile stretch of land. The combination of the two factors might also be a possible explanation. The author feels that the story, on the whole, is based on a historical event, namely, that during the period when the tribes of Israel advanced into the Land of Canaan west of the Jordan, the river was rather low following a series of dry years. During the same period a series of earthquakes occurred, causing obstructions which stopped the flow, thus giving warring tribes the opportunity to cross the Jordan in places other than the usual fords which were most probably guarded by the people of the cities nearby. The reason for maintaining that this event did occur is the preciseness of the hydrological description of this event. "That the waters which came down from above stood and rose up as one heap very far at the City Adam, that is beside Zarethem, and those that came down toward the Sea of the Arava, the Salt Sea, failed and were cut off and the people passed over right against Jericho" (Joshua 3:16). City Adam is the place where such a dam would form, the effect of which would terminate at the Sea of Salt. It might be that the same earthquake caused the walls of Jericho to crumble, enabling the people who crossed the Jordan to storm the city without having to scale the walls.

Jericho was the first city which stood in the way of the Israelites. The reader has already some background to the history of this ancient town. From the point of view of compatibility with the archeological records, there is a difficulty for the biblical archeologists. The story of the destruction of Jericho by Joshua and his Israeli army does not correspond with the results of the archeological excavations of the Tel of Jericho. No remnants of a destroyed, large town of the Late Bronze Age were found in the excavations carried out [1]. The author's opinion is that, as in general, evidence of the conquest has been found in many tels and as only a relatively small part of Jericho has been excavated, there is still a good chance that a small walled town will be found in the large, as yet unexcavated, Tel of Jericho. It is difficult to accept the idea that the site of Jericho, in the close vicinity of the big spring, was abandoned and did not form an obstacle to the invading Israelites. Thus, without further comment on the question already raised about the accuracy of the biblical story, the general story of the fording of the Jordan and the conquest and destruction of the city seems, to the author, fairly authentic, in view of its conformity with the general geographical and hydrological set-up.

As mentioned earlier, the general story of the conquest of the Land of Canaan is in concordance with the archeological records. In many Canaanite towns the conquest can be traced by a typical layer of ashes, the dating of which gives a time of about 1200 years B.C. Overlying this layer is another, characterized by a new type of ceramics, which the archeologists attribute to the Israeli period.

For some ruins, such as those attributed to the ancient city of Ai, the archeological record does not conform with the biblical story. In such cases, the reason may be the incorrect identification of the site by the archeologists or inaccuracy in the Biblical record. As previously suggested, one should not expect the Bible to be a precise and scientific record of history, but a general description of the historical events and environmental background including the religious and subjective interpretations of its observers. On the other hand, it is not suggested to take the conventional explanations by the biblical archeologists as an absolute truth. These change as research advances and more data is added from excavations. On the whole they should be taken as a working hypothesis, best explaining the up-to-date findings. To the best knowledge of the present author, severe changes in the climate were not taken into consideration by most of the archeologists, and the author takes the liberty of suggesting alternative scenarios. In addition, one has to take into consideration the likelihood of reinterpretation and additions which might have occurred during the retelling or recopying of the stories by each successive generation. This happened until the story was finally put together. This probably started at the time of the First Temple when King Solomon established a central bureaucracy of official scribes similar to that of Egypt. The scribes put into form and style all the pieces of verbal and written information they were able to gather and store in the archives of the Temple [2]. The final codification was most probably done even later.

In some of the stories of the conquest, the reader can find an apparent contradiction. An example is the cities of Jerusalem and Hazor, which are included in the list of cities conquered by Joshua (Joshua 12), yet later, in the time of the Judges, they are reported to be flourishing Canaanite towns. The archeologists suggest that many of the conquests of later periods were attributed to Joshua's army which had made the first assault into the land, broken the opposition of local inhabitants in the mountains and foothills, and penetrated the northern parts of the country. The whole process of conquest may have taken about 200 years. The flat lands, as told in the Bible, were avoided by the tribes who could not stand against the "iron chariots" of the Canaanites. "And the children of Joseph said, the hill is not enough for us and all the Canaanites that dwell in the land of the valley have chariots of iron, both they who are of Beth-Shean and her towns and they who are of the Valley of Jezreel" (Joshua 17:16). As we will see later, the Israelis found a way to cope with these chariots but it is quite obvious that, for a while, this weapon halted the onslaught by the Tribes of Israel.

Many of the Canaanite towns, such as Hazor, Gezer, Megiddo, Gibeon, and Jerusalem, had another "secret weapon" which enabled them to withstand besieging armies. Each of these towns had a secret approach to a spring below the city so that, in time of war, the people still had access to a supply of water without the danger of having to leave the towns. After conquering these towns, the Israelites adopted and developed this system into one of the wonders of ancient engineering, yet the invention of the system has in all probability to be attributed to the Canaanites, who were the first to build their towns on the hills above the springs.

In some of the towns, the system was not built in one stage but was developed, step by step, either as a function of the development of the town or the engineering faculties of its people.

The system in the town of Gibeon can be taken as an example. This town is built above a spring issuing from the limestone rocks of Middle Cretaceous age. As told in the Bible, the people of Gibeon outwitted Joshua when they surrendered peacefully and had his oath to save them from annihilation. On discovering the deception, Joshua made the Gibeonites into "hewers of wood and drawers of water for the (Israelite) congregation." These people could indeed teach others how to draw water as in their city a very big shaft and spiral staircase had been cut into the limestone rock so the water carriers could walk down to the water (Fig.10.2). At the bottom of the shaft are two small caves in which some water is found. Although the system is dry today, it was undoubtedly originally excavated to locate a perched water table, the source of a small spring outside the town. The existence of such a spring becomes obvious from the other water project built in the near vicinity, which comprises a tunnel and stairs built on two levels with two gradients. This meant that the first stage was the construction of a tunnel starting at the walls of the city and reaching the level of a small spring outside the town. When an enemy approached, the outlet of the spring was blocked and the people could have access to the water via the tunnel. At some stage the upper spring dried up, in all probability because the town expanded over all its replenishment area. The tunnel was then extended to reach the level of the lower spring, which has a larger replenishment area and to this day is flowing steadily [3]. It also seems quite possible that the preliminary stage was the driving of a tunnel from the outlet of the spring into the aquiferous rocks in order to increase the quantity of the water of the springs, particularly at the end of the summer. Water from this spring was then and still is being used for irrigation.

From the excavation of the tunnel, the people learnt that the water table was continuous below their town. This gave them the idea to drive a tunnel from the town to meet the lower water table in order to secure a water supply for their city in time of siege.

Fig. 10.2. The big cistern of Gibeon (Photo courtesy Irit Zaharoni Bamahne Journal, IDF)

The big shaft with the spiral staircase at Gibeon has been interpreted by archeologists as a water supply project. The present author's opinion is that it was a site of a cult to a water deity. Moreover, it is argued that the name Gibeon is in some way a later change of the original Canaanite name "Gaevon" where "gaev" means water hole or shaft and "on" is the usual Canaanite Semitic addition which denotes strength and power, as in Dagon, Shimshon, Hermon. The present name of the village, A-Jib, which in Arabic means "water hole", may provide evidence of this. If the present author's suggestion, that the big shaft at Gibeon was a place where a local water deity represented by the spring was worshipped is accepted, then the big, wide staircase leading to a very small spring can be explained as the ceremonial way for a procession of priests to go down to take some water from the dripping cave wall to be used as a ceremonial offering on the altar in the local sanctuary. That such a cult did exist in the Canaanite towns can be concluded from celebrations connected with the libation of water on the altar exercised by the Hebrews who replaced the Canaanite inhabitants. The most famous example is that of the "Celebration of the Place of the Watering" which took place in the Temple of Jerusalem on the holiday of Succot, in the autumn, when the High Priest would bring water from the Spring of Gihon. That this cult was of Canaanite origin can be deduced from the fact that nowhere in the five books is it mentioned, although a complete list of daily and yearly sacrifices is methodically disclosed in the books. (a more detailed description of this water libation is given in Chapter 12).

As the story of the water project of Jerusalem will be discussed later when the Period of the Kings is reached, it is suggested we return to the time of the conquest of Canaan and see more of the role of water in the fate and faith of the people of that time.

Before leaving Gibeon, it should be mentioned that this big shaft of Gibeon is mentioned several times in the Bible, as it was quite a famous landmark. In the later period (Samuel II 2:12-17) it is called Brecha, namely, cistern or water reservoir. It seems possible that the local cult stopped at a rather early time, the main reason being that the upper water level of the spring dried up. It was most probably used afterwards as a temporary reservoir to collect some of the runoff rainwater from the surrounding area. Yet it does not seem to the author that it was originally dug as a water reservoir, as the rock is very permeable and the covering of all the walls with plaster does not seem feasible.

The special device for ensuring a supply of water in time of siege was not used by the Gibeonites, who surrendered immediately, but did help other Canaanite cities to withstand the first onslaught of the Hebrews. Archeological excavations in some of the cities which are listed in the Book of Judges as not being conquered by the Hebrews, including Gezer, Jebus or Jerusalem, Megiddo, Ibleam, and Hazor, each had a system of tunnels to supply the city with water in times of siege. This is an opportunity to describe these systems which most probably played an important role in the resistance of the Canaanites to the onslaught by the Israelites and, as we will see later, also played an important role in the history of Jerusalem.

The reader has already been made aware of the fact that most Canaanite towns were built near a source of water, in most cases a spring, in some places, wells. In later periods the art of collection of rainwater in artificially plastered, watertight cisterns was developed, but during the Canaanite periods this method was restricted to areas where the rock layers were naturally impermeable.

As time progressed the tels on which the towns were built rose steadily above the springs as the result of each generation's building on the ruins of the dwellings of former generations. The growing tel had the advantage of greater security for the people of the town, especially once they learned to build heavy walls on the summit of the tell. On the other hand, these towns became less secure in times of siege as they were cut off from the spring or well which provided their supply of water. Some time during the Canaanite period, in one of the towns the people conceived the idea of excavating a tunnel from the top of the tel to the spring. Prof. Ruth Amiran of the Hebrew University of Jerusalem has suggested that the Canaanites adopted this method from the Mycenaean in Greece. She points out that in Athens and in Mycenae two such tunnels led from the Acropolis of the towns to a spring at the base of the towns. Indeed, a Mycenaean bowl was found in the tunnel of Gezer.

Due to the fact that the excavation of a tunnel does not leave any building remains it is difficult to date, except in a case where early ceramics are found, as in Gezer. This is the reason that the archeologists are divided in their opinion regarding the time of the excavation of the tunnels. Some maintain that the tunnels of Megiddo, Jerusalem, and Gibeon were built by the Israelites, others that they were excavated earlier, during the Late Bronze Age, between 1425 B.C. and 1200 B.C. On the basis of hydrological considerations, the present author thinks that the earlier age of the tunnels seems more feasible. His reason for this is that in most of these systems one finds that the upper part leads to an upper dried up system, which seems to be a small spring which was flowing before the city was built [3]. Another consideration is that one finds water tunnels in all the Canaanite cities like Megiddo, Hazor, Gezer, Ibleam and Jerusalem, which resisted the Israelite conquest (Judges 1). This does not necessarily mean that the existence or absence of such a tunnel was the reason why a certain town was conquered or not, but it is worthwhile noticing that all the towns (except Gibeon, which surrendered) which are known for sure to have had such a water tunnel are included in the list of unconquered Canaanite cities listed in the first chapter of the Book of Judges.

Other types of water projects were very deep shafts excavated from the top of the tel to the water table, such as one finds in Tel Lachish, or inclined shafts like that of Gezer. In both cases, and also in Hazor, one can see that the planners might have had an idea about the existence of a groundwater table and extended their tunnel to the main spring only after the local perched water table dried up or was not sufficient because of the covering of its recharge area by the buildings of the expanding city. In the Bible, the only place for which the excavation of such a tunnel is mentioned is Jerusalem, which will be dealt with in more detail later on.

Another method of securing water for the town dwellers was the collection of rainwater in storage cisterns. This method could be applied only in areas built of impermeable rocks which do not allow leakage of the water stored in the cisterns. Indeed, in the mountains of Shomron (Samaria) which are built of impermeable chalks, one finds early Canaanite villages and towns (18th-17th century B.C.), which had no spring in their vicinity. These places undoubtedly adopted the solution of rainwater storage. During the late Canaanite period (16th-13th century B.C.), one finds the expansion of Canaanite villages onto the permeable limestone areas. Thus one may conclude that the inhabitants of these places had found the secret of waterproofing their cisterns using impermeable plaster.

The water supplied by collected rainwater could be enough for a village or small town; however, it would not be a secure supply for a large town for a long period of siege during the months of a long summer, especially when preceded by a dry winter. Thus, one finds all the big fortified cities near springs, some of them utilizing the device of the tunnel to provide a safe approach to the water supply even in times of war. During the later periods, the people developed the dual system which was built on both storage of water in cisterns and the supply from the springs.

When the Tribes of Israel started their conquest of the Land of Canaan, they had to cope with problems which were new to them against which they had no tools and experience. The first obstacle was the fortified cities backed by sophisticated water supply systems; the second was the armored chariots (called "iron chariots", though most probably they were clad in bronze) with which the Canaanites were equipped and which enabled them to storm the Israelites who were besieging their towns or fighting in the open field. Thus the Israelites had no alternative but to settle in the mountainous areas where no big cities existed, but which were not suitable for armored chariot warfare, and to adopt the method of cistern calking.

The problem they faced in the mountains was that of withstanding severe erosion of the soils after the woods had been cleared. The mountainous topography and the Mediterranean-type climate of heavy rains after long dry summers would cause the soil from any field cleared of trees to be swept into the river-beds and from there to the sea. Thus when the tribe of Joseph complain to Joshua that the lot which was given to them is too small and they cannot drive out the Canaanites from the plains because of their "iron chariots", Joshua advises them to go to the mountains and cut down the woods (Joshua 17:15). The Bible does not mention the need for the application of any special agricultural method to overcome the problem of erosion. Thus, it might be that the first settlers, like most nomads coming from the desert, wanted the land for their sheep and goats and did not care much about soil erosion. Yet, after living some time in this area and witnessing the soil being washed away from their fields, and seeing that many of the Canaanite towns survived, they slowly learnt from their neighbors the art of terracing. Gradually, subsequent generations cut down most of the natural vegetation and replaced it with fruit trees such as grapevines, figs, carob and olive trees.

When did the Canaanites begin to settle the mountains and adopt the terracing method? To this question we have no written answer, yet the archeological evidence from early Canaanite settlements in the mountain areas and the pollen assemblage from the core samples taken from the bottom of the Sea of Galilee [where at around 5000 B.P. (3000 B.C.) one can see a severe reduction in the number of oak pollen and an increase in the number of olive pollen(Fig.1.3)], suggests that the cutting down of the natural forest of oaks and its replacement mainly by olive trees started during the Early Bronze Age. In the same core, some time between 3100 and 3300 B.P., which is the period of Israelite settlement, one observes a small increase in the olive pollen and a reduction in a certain type of oak (Thaburensis), as well as grain legumes. Is this the marker for the settlement by the Israelites combining wood cutting, overgrazing, and the gradual building of new terraces and planting of fruit trees, mainly olives?

The settlement of the Tribes of Israel in the Land of Canaan was most probably a slow process of penetration, wars, and the reclamation of the wooded noninhabited hills and mountains. It is logical to assume that, from the economic point of view, in the beginning the Israelites depended mainly on their livestock, but slowly adopted the agricultural methods of the local inhabitants, especially the methods of terracing the hill and mountain slopes, diverting water from springs for their irrigation, and building cisterns to collect rainwater.

Their living together with the local population and the need to learn their agricultural tradition also brought about the adoption of the fertility rites of Baal, Asherah, and Ashtoreth, which were, for the Canaanites, an all-inclusive culture. This brought in its wake the assimilation of the peasant population with the local population, which had its cultural-religious implications for many years to come.

The wars of settlement continued after the death of Joshua, into the period of the Judges. One of the famous battles was that against the Canaanite army of armored chariots commanded by Sisera, the captain of Jabin the King of Hazor. This battle reveals the process of settlement. In the first place, it can be seen that, contrary to the report about the conquest of Hazor by Joshua (Joshua 12:19), this town was at the center of a flourishing Canaanite kingdom at the time of Judges. The archeological evidence from the excavations at Hazor support the Biblical story about the city's being conquered and destroyed by the Israelites some time in the second half of the 13th century B.C., yet one cannot define the time of the conquest exactly. The conclusion is again the general one, that in the Bible we find the general story of conquest and a general picture of the social, cultural and historical background, while the detailed story has to be put together, piecemeal, by the use of archeological, historical, geographical, and other pertinent data. Thus the historical order of events was that in the beginning the army of the King of Hazor might have been beaten by Joshua. The unwalled part of the city might have been sacked, while the walled part was able to withstand the attack and regain sovereignty over the plain of the upper Jordan. The city's ability to withstand attack would probably be helped by the anti-siege water system with which the city was provided at that time.

The second battle, at the time of Deborah, was in the open field. The Hebrews were oppressed by the Canaanites and could not fight back, as they lacked the armored chariots which were the Canaanites' main corps. A coalition of the tribes, headed by Deborah, chose Barak, the son of Abinoam, as their leader to try and break the yoke of the Canaanites. Barak, whose name means "lightning", outmaneuvered the chariots of the Canaanites by drawing them after him to the banks of the stream of Kishon in the midst of the plain of Jezreel, which is famous even today for its marshes and swampy soil, especially during winter. A sudden torrential downpour, which caused the flooding of the banks, helped Barak accomplish this maneuver. The chariots sank and were held by the heavy black mud of the bank and some of them were even swept into the flowing water. Barak's infantry warriors were quick to attack the chariots, which had lost their main advantage of mobility, causing Sisera to flee on foot in search of refuge in the tent of Yael the Kenite, where he was given eternal refuge.

Another battle took place in the Valley of Jezreel, this time with a hydrological prelude. This was Gideon's fight against the Midianites, the desert nomads who took advantage of the lack of a strong central government to prevent them from overpowering and impoverishing the people by sacking and pillaging. Gideon's first move was to reduce the number of people who assembled for the war, to a small fighting army. He brought the army to the spring of Harod (Fig. 10.3) in order to observe the behavior of his warriors. This rather small spring emerges from the limestone rocks of the Eocene age which form the upfaulted block of Mount Gilboa. Those warriors who showed a touch of refinement by kneeling and drinking from the cup of their hands were sent home. The real field men who stretched on their bellies to lap the water as a dog does, a sign that they knew how the hunter moves silently and never takes his hand off his weapon, were chosen for the night assault on the huge Midianite camp. The assault came as such a surprise to the Midianites that they panicked, killed each other in the darkness, and fled over the fords of the Jordan, back to the desert.

At the end of the period of the Judges, the Canaanites were either subjugated or assimilated. At this time a new enemy, the Philistines, arose to harass the Hebrews. These "new Philistines", were not the descendants of the so-called "Philistines" of the time of the Patriarchs. The new people, known also as the "People of the Sea", were newcomers to this region.

The Philistines conquered the coastal plain at the same time that the Hebrews conquered the foothills and mountains. As war was a way of life for them, they proved to be a hard enemy to deal with. They brought with them methods of warfare and new types of iron weapons which gave them an advantage over the Hebrews whose traditions and weapons had not changed since the Late Bronze Age. However, the main wars between the two nations took place at the beginning of the period of the Kings. In fact, one gains the impression that the need to fight such a strong and organized enemy as the Philistines brought the tribes of Israel to the conclusion that they could not go on fighting tribe by tribe, not to speak of tribe against tribe, but needed a central government which would concentrate and organize their forces to withstand the onslaught from their old and new neighbors.

The war between the Philistines and the Hebrews continued during the reign of King Saul, the Israelites fighting a defensive war against the invading neighbors. The war reached its peak in the battle of Gilboa, where King Saul had assembled all the tribes to try and repulse the invading army. He failed and paid for this with his and his sons' lives. A dissident outlaw, by the name of David, the son of Yishai, who had killed the Philistinian hero, Goliath, and later served as a mercenary to the Philistine tiran of Gat, Achish (The Achaian?), was elected chief of the tribes of Judah and Benjamin and later, king of all the tribes of Israel. David's first move was to conquer the citadel of Jerusalem from the Jebusites and proclaim this city the capital of all the tribes of Israel.

Fig. 10.3. The Spring of Harod (Photo by the author)

We know of the arrival of the Philistines at this part of the Mediterranean coast from Egyptian sources. In the eighth year of Ramses the Third, who reigned from about 1200 B.C. (first ruler of the 20th Dynasty), he describes in detail his wars and triumphs over the "People of the Sea", among them, the Peleset tribe. In an earlier report, King Merneptah (of the 19th Dynasty), who ruled from 1224 to 1214 B.C., described his war against the Libyans and the People of the Sea. In the same stele the same Merneptah declared that he had destroyed Israel forever. On the other hand, in neither the El-Amarna letters found in the Archives of Amenhotep (Amenophis) III (1398-1361 B.C.) and Amenophis IV (1361-1353 B.C.), the latter also known as Akhenaten, nor the triumphal declarations of Seti I (1302-1290 B.C.) and Ramses II (1290-1224 B.C.), who reached the northern border of Canaan, are "The People of the Sea", among them the Philistines, yet referred to as an enemy to be dealt with.

The putting together of all the evidence, namely, the written and illustrated Egyptian records, the stories of the Bible, and also archeological findings, enables the formation of a picture which, though not fully detailed, gives the general impression of the period of the settling of the Philistines and their wars with the Israelites.

The area in which the Peleset tribes settled after their defeat by Ramses III was the southern coastal plain of Canaan, extending to the east to the foothills of the mountains of Hebron. Although, as far back as the time of the Patriarchs, the inhabitants of this area are called Philistines by the Bible, it is clear that the reference is by later writers who gave them the name by which they were known to them. It is quite remarkable that the King of the Philistines of the time of the Patriarchs had the Semitic name of Abimelech. It is thus quite probable that these Semites were conquered, expelled from the coastal plain, and the area inhabited by these People of the Sea.

A present-day space picture of the region (Figs. 1.1,4.1) shows us that most of the area is covered by sand dunes.These sands are young and started being deposited at the time of the conquest by the Arabs about 660 A.D. At the time when the Philistines settled there, as the investigations of the present author have shown [4], this area was covered by reddish sandy clayey soils (Hammra) produced by the decomposition of the calcareous sandstone of Pleistocene age which form the bedrock in the Coastal Plain. The sandstones were formed by the consolidation of the quartz sands which were brought to the land by the waves and currents of the sea, mainly during interglacial periods. The sources of the sands, as already mentioned, are in Nubia, from where they are carried by the Nile. The consolidation of the sands is mainly by carbonates of marine origin, spread by waves during heavy storms, or airborne, but also through the dissolution of carbonate shells by rainwater and precipitation by capillary ascent and evaporation. The clayey component of the soil is, in the opinion of the author,due to the dust brought by storms during the humid (glacial-pluvial) periods. The subsurface is built of alternating layers of sandstones and clays, a result of the climatic changes which affected this region during the Pleistocene. The regime of transgressions and regressions brought about the formation of longitudinal ridges of sandstones, running parallel to the coast, each ridge representing an ancient shore line during one of the interglacial periods. (The coastal ridges of the regression phases are submarine). The continental ridges formed obstructions to the flow of the rivers to the sea, causing swamps to be formed behind them in which clays were deposited.

The local red soil (the reddish color is due to the weathering of the heavy minerals brought with the sands) has the right ratio between sand and clay; as the region is close to the sea it receives enough rains to make it a very fertile land. Only in the longitudinal valleys where clays accumulated do drainage problems exist. That part of the rainwater which does not evapotranspirate, infiltrates the subsurface, accumulates on the underlying clay layers, especially that of Neogene age (which underlies all the coastal plain) to form a regional groundwater table (Fig.10.4). The groundwater flow is toward the sea . Thus, near the seashore the water table is very near the surface and can be reached by shallow wells. Moreover, the ephemeral streams which flow through the area reach the groundwater table near the shore and become perennial by draining the water table. This ensured a supply of water to the cities located along the shore. Due to this reason, the main road from Egypt northward was along the shoreline ("The way of the land of the Philistines" Exodus 16:18). The Philistine inland fortified cities, like Gat, were situated at the contact line between the chalk-built area of the foothills (Eocene age) and the marl-built area (Neogene age) of the western coastal plain. This caused the formation of a north to south line of springs, along which settlements had been situated since the time of the Early Bronze Age.

The road along the shore served the Egyptian kings in their conquest campaigns of the Land of Canaan. The kings of the 18th Dynasty after expelling the Hyksos, extended their dominion over all Canaan and even reached Mesopotamia. The road continued to serve during the reign of the 19th Dynasty, from which time there exists a map of King Seti II (ca. 1200 B.C.), one of the last of the Pharaohs of the 19th Dynasty, which portrays the road from Egypt to Canaan along which a series of Egyptian fortresses and garrisons are schematically described (Fig. 8.2). In this, the dominion of the Philistines over this area is not yet mentioned, though they might have already settled there by that time.

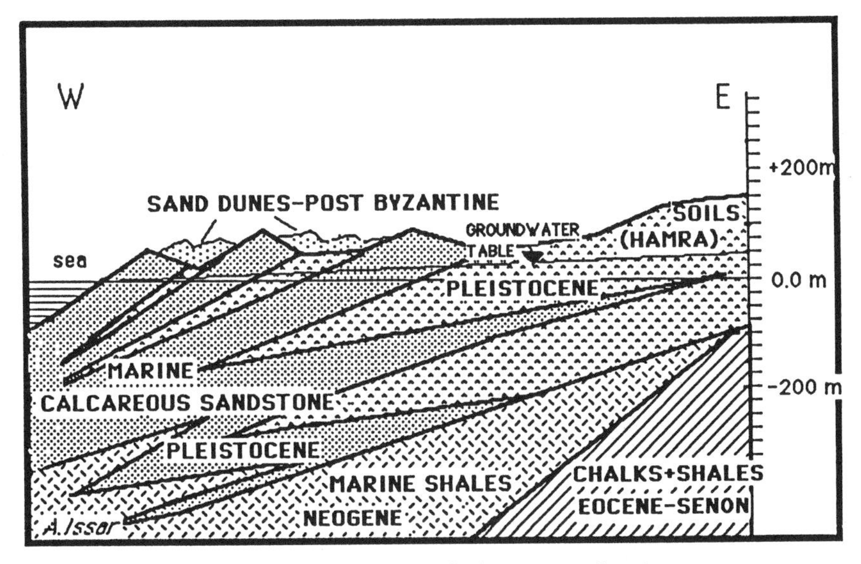

Fig. 10.4. A geological cross-section of the Coastal Plain, southern Israel

The archeological excavations which can be correlated with the stories in the book of Judges show that the establishment, expansion and wars with the Israelites were between the years 1100 B.C. and 1000 B.C. During that time it looks as though the Egyptian kings were no longer the masters of the "Way of Horus", as the Egyptians called the line of fortresses and garrisons, but that it was an independent Land of the Philistines. At that time these people defeated the Israelites and conquered most of their land west of the Jordan, reaching as far as Beth-Shean and taking the Holy Ark into captivity, until they were conquered and subjugated by King David and his dynasty. But before they vanished from the stage of history, to leave only their name to this corner of the Levant, the events which caused their appearance and expansion were very much connected with the fate of the Israelites.

The starting point of the chain of events which, at its end, brought about the collision between the Philistines and the Israelites, might have been also the environmental change caused by the huge explosion of the cycladtic Island of Thera (Santorini) in the Aegean Sea, which happened at about 1600 B.C. The immediate impact of this explosion caused a series of earthquakes, tsunamis (destructive tidal waves), and fires. This caused the destruction of the Minoan empire of Crete and its capital city, Knossos. The hegemony of the Minoan civilization passed to Mycenae on the mainland of Greece.

The explosion of Thera, which formed a crater about four times that of Karakatao, undoubtedly poured tremendous quantities of volcanic ash and dust into the atmosphere, which most probably caused a reduction in solar radiation and a global cold wave. It may have been this cold spell which triggered the migration southward of tribes from the Carpatho-Danubian plains. As already mentioned, the same cold event, on the southern coast of the Mediterranean probably also had a causal effect on the Exodus of the Hebrew tribes from Egypt. In the northern Mediterranean, it apparently produced a chain of migration toward the south. When the Hebrew tribes were conquering the mountains and hills of Canaan the waves of people from the north reached the shores of Egypt. The first wave came during Pharaoh Merneptah's reign, at about 1220 B.C. The second, which was most probably the crest of the migration, was at about 1200 B.C. at the time of Ramses III. The Peleset or Philistines, defeated by the Egyptians, then settled in the coastal plain, probably with the consent of the Egyptians with whom they came to an agreement.

The Philistine settlements are filled with Mycenean pottery of post-Minoan period (IIIC) dated to about 1200 B.C. [5], which conforms with the dates of the Egyptian sources. They also brought with them the Mycenaean system of warfare which was based on towns headed by "Seranum", a probable equivalent to the Greek "tirans", bonded by a treaty in time of war. Their armaments were described in the Bible in reference to Goliath and are illustrated on the "Warrior Vase" found at Mycenae, dated 1200 B.C. The warriors wore helmets, coats of mail, and greaves of brass or bronze, and were armed with an iron-headed javelin and protected by a large round shield.

The illustration of Ramses III and the battle between David and Goliath show the Philistines as champions of close-combat with sword, javelin and spear, while the Egyptians preferred longer-range combat with bow and arrow. David used the sling, which had an even greater range. Yet, in the first stages of the war, the Philistines had the upper hand. They managed to defeat the Israelites in a series of combats, capture the Holy Ark and bring it to the temple of their god, Dagon, who, as mentioned earlier, was most probably of Semitic origin. His name may be connected with corn (Dagan) or with the fish (Dag). He was the god of fertility of the soil and successful corn yields, a product most important to the Philistines, not only for food, by also for their beer production. Beer, as evidenced by the many special jars with sieve-mouths, was an important part of their culture.

In order to combat the Philistines, the Israelites assembled under one judge, the prophet Samuel, and managed in one battle to repulse their enemies, but the freedom was not complete. The Philistines held garrisons throughout the land and oppressed the Israelites. As the bond between the Israelite tribes was unstable under a judge, they urged Samuel to appoint a king at their head in order to fight the Philistines. Samuel did not like the idea of a monarchy. He gathered the people at Gilgal and tried to convince the tribes that it was not worthwhile to change from a democratic system with an elected judge to an inherited kingship. He listed all the positive sides of an elected judge against a king, yet he was not able to convince them. He tried to use a meteorological anomaly of a thunder and rainstorm in the time of the wheat harvest to convince them, yet the people insisted on a king, and Samuel, after warning them against following the evil route, gave his consent, and Saul from the tribe of Benjamin was elected as the first king of the tribes of Israel.

The first thing the new king did was to attack the Philistine garrison at Geba. This was a proclamation of war and the Philistines were quick to respond. They collected their army of chariots, horsemen, and foot soldiers at Michmash. Many of the Hebrews panicked and hid themselves in "the caves, thickets, rocks, high places, and pits; and some of the Hebrews went over the Jordan to the Land of Gad and Gilead" (Samuel I 13:6,7). Saul waited for Samuel to come and perform the holy ceremony of sacrifice before he went to war. As Samuel did not join him and the army began to disperse, Saul carried out the holy ceremony himself, which undoubtedly irritated old Samuel, who had started to plot against Saul. By a brilliant, unexpected move, Jonathan the son of Saul managed to beat back the Philistines.

The war between the Hebrews and the Philistines continued and in one of these battles a ruddy young shepherd boy named David excelled by using the sling against an armored Philistine. This gave him some ideas how to fight this enemy. When King Saul and his son Jonathan were killed in the battle of Gilboa, and the Philistines were again in control, he lamented the fall of Saul and opened his lament with the bid to "teach the children of Judah the use of the bow, behold it is written in the book of Jashar" (Samuel II 1:17). In other words, to adopt the tactics of long-range combat. In his lament he curses the mountains of Gilboa in the semi-arid eastern part of the Jezreel Valley for being the battlefield on which the kings of the Hebrews were killed. "Ye Mountains of Gilboa, let there be no dew, neither let there be rain upon you, nor fields of offerings; for there the shield of the mighty was defiled, the shield of Saul, as though he had not been anointed with oil" (Samuel II, 1:21).

After a war with the followers of the house of Saul led by Abner the son of Ner, David triumphed. One of the skirmishes was at the water shaft or pool of Gibeon (Fig. 10.2) discussed in the previous chapter. " And Joab the son of Zeruiah and the servants of David went out and met together by the pool of Gibeon: and they sat down, the one on one side of the pool and the other on the other side of the pool. And Abner said to Joab, let the young men now arise and play before us. And Joab said, let them arise. Then they arose and went by the number twelve of Benjamin...and twelve of the servants of David and they caught every one his fellow by the head and thrust his sword in his fellow's side, so they fell down together" (Samuel II 2:13-17). In this skirmish, as well as in other combats, Joab the son of Zeruiah, David's army leader, triumphed. When Abner the son of Ner was killed by Joab, the followers of the house of Saul panicked. The last heir, the son of Jonathan, was betrayed and killed, and his head was brought to David, who was camping in Hebron. David did not pay the killers, as they most probably had expected, but "slew them, cut off their heads and feet, and hanged them up over the pool in Hebron" (Samuel II 4:12). This pool, which is a large reservoir into which rainwater collects is still in use in the town of Hebron, though no longer for the town drinking water nor for hanging the heads, feet, or any other remains of the bodies of traitors above it. It is now mainly used as a source of drinking water for the herds of goats and sheep during the winter and spring.

11 Kings, Tunnels, and Canals

> And the rest of the acts of Hezekiah, and all his might, and how he made a pool, and a conduit, and brought water into the city, are they not written in the book of chronicles of the kings of Judah? (Kings II 20:20)

> And when Hezekiah saw that Sennacherib was come, and that he was purposed to fight against Jerusalem, he took counsel with his princes and mighty men to stop the waters of the fountains which were without the city, and they did help him. So there was gathered much people together, who stopped all the fountains, and the brook that ran through the midst of the land, saying, why should the kings of Assyria come and find much water? (Chronicles II 32:2,3,4)

> This is the story of the boring through: whilst (the tunnelers lifted) the pick each towards his fellow and whilst three cubits to (be) bored (there was heard) the voice of a man calling his fellow, for there was a way in the rock on the right hand and on (the left): And on the day of the boring through, the tunnelers struck each in the direction of his fellows, pick against pick, And the water started to flow from the source to the pool twelve hundred cubits. A hundred cubits was the height of the rock above the head of the tunnelers. [The Niqba inscription] (Fig. 11.4)

King David's conquest of Jerusalem seems to have been an extraordinary deed. The Jebusite citadel survived conquest by the Israelites. Although the Amorite King of Jerusalem, heading a coalition of Amorite kings, was defeated in the field, the City of Jerusalem was not conquered. It defied even King Saul, whose main city was in Gibeah, about five kilometers to the north. One of the reasons that the Jebusites could withstand the assaults of the Israelite warriors was the topographical set up of their city. It was built on a hill surrounded by steep gorges on three sides. The fourth side was probably defended by a wall and a trench. Another reason might have been the water supply which enabled the city to withstand siege. One deep shaft was excavated from within the city walls to reach the water table of the spring of Gihon ("the emerging one").

The Bible tells us that the Jebusites ridiculed David's attempts to storm their citadel and called to him that even their invalids would defy his army. David then offered a prize, " whosoever smiteth the Jebusites and getteth to the gutter" (Samuel II 5:8), which Joab his commander won by a special feat which is not very clearly described in the Bible.

After the ancient water supply system was found and excavated by the British archeologist, Captain Charles Warren, in 1867, there was a general tendency among the bibliologists to explain this chapter as involving this ancient system. Their opinion is that Joab discovered the shaft which connected the Jebusite town to its water resource and, by climbing it secretly, was able to enter the town without having to storm the walls. This explanation seems feasible and shows the vulnerability of such a system. Should the enemy discover it, he could either cut off the water supply, or even use the system to enter the town and attack from within, although it seems that it would be very difficult to gain entry to any town through such a shaft. The specific mention of the "gutter" in the case of the conquest of Jerusalem and the discovery of an elaborate system of shafts and tunnels speaks for the conclusion that David's conquest of Jerusalem had to do with its ancient water system. Although some archeologists think that the system which was discovered by Warren is Israelite, the present author's opinion, as will be explained later, is that it was initiated by the pre-Israelite inhabitants of the city.

After David secured control over Jerusalem, he expanded his rule beyond that which the tribes of Israel had already done. He subjugated the Philistines and established a small empire which extended from Syria to the borders of Egypt.

We are told that during the fight against the garrison of Philistines at Bethlehem in Judea, three of David's warriors broke through the Philistine's lines and brought him water from the well in Bethlehem. David would not drink this water, as it was brought in jeopardy of the warrior's lives. Instead he "poured it as a libation unto the Lord" (Samuel II 23:16). This, and the occasion when the judge and prophet, Samuel, "poured water before the Lord" (Samuel I 7:6) are the only places in the Bible where libation of water is mentioned. One can thus assume that the libation of water was practiced by the Israelites though not to the extent and with the ceremony it developed during the time of the Second Temple.

David's successor, Solomon, was a king of peace and construction. Although he did not war against his neighbors, he prepared his country for times of war by fortifying many towns. The fortifications were according to a general plan which included walls, a fortified gate, and a water supply system. The plans made by the royal architects were followed so exactly by the masons that once the archeologist Yigael Yadin became acquainted with them, he could tell his workers exactly where to dig in order to unearth what remains of these constructions.

Fig. 11.1. The staircase leading to the water supply tunnel at Hazor (Air photo courtesy Irit Zaharoni, Bamahne Journal, IDF)

Some of these fortifications were destroyed during the reign of Solomon's son, Rehoboam, by Pharaoh Shishak (Sheshonk), who crossed the Land of Judea and Israel, destroying and pillaging many cities. Some of these towns were rebuilt and refortified by the later kings of Judea and Israel. There was a period of elaborate building and fortifications during the reigns of the Israelite king, Omri, and his son, Ahab.

The water works furnished for these towns by King Ahab's engineers are remarkable. They did not invent these systems, but developed them to the extent that they could serve a large population taking shelter in a town for a long period of siege. By building these systems, they showed that they had a basic knowledge of groundwater flow. This concerns an understanding of the water table phenomenon. In other words, the Israelite engineers knew that if springs are found in the vicinity it was not necessary to reach the spring itself in order to get to its water but it was sufficient to go down to the topographical level of the spring and touch the layer which supplied its water.

Thus the water systems of Hazor and Megiddo do not even go in the direction of the springs emerging below the cities, but to the water tables feeding them. It is quite possible that this knowledge was acquired from the Canaanites, who formerly ruled these cities, yet King Ahab's engineers perfected this knowledge into impressive water supply projects.

At Hazor, the system was built with a very wide staircase going in a gradual gradient from the center of the town to a tunnel where the water table is found (Fig. 11.1). At Megiddo, the system is composed of two parts: a long tunnel which might have been an ancient excavation carried out to increase the flow of the springs; and a declining shaft and staircase to reach the tunnel (Fig. 11.2).

Fig. 11.2. Reconstructed model of Megiddo staircase to the water supply system (Photo courtesy Irit Zaharoni, Bamahne Journal, IDF)

Going back to King Solomon, he was very keen on promoting his international relations. He concluded a pact with the Canaanite King of Lebanon, Hiram of Tyre, in which he surrendered a certain territory in order to obtain cedars to build the Temple in Jerusalem. He established a port on the Red Sea and sent a mission to Ophir (probably Africa) to bring exotic merchandise, in the wake of which the Queen of Sheba visited Jerusalem. Friendly diplomatic relations with the court of Egypt were also established and a daughter from this court was given to the King of Jerusalem in marriage. The King of Egypt, though a friend to the King of Judea and

Israel on the diplomatic level, still held memories of the times when Canaan was under the domination of the Egyptian kings. Thus, when a political refugee from Israel by the name of Jeroboam asked for asylum in his country, it was not denied him. As an heir to an ancient line of rulers, the Egyptian king already knew the proverb, later to be phrased in Latin, "divide et impera". When King Solomon died (about 930 B.C.) the people of Israel requested his son, King Rehoboam, to lighten the yoke of taxes his father had levied from them for the purpose of carrying out his extravagant projects, but the heir refused. This was the occasion for Jeroboam to call for a rebellion of the tribes of Israel and separate them from the House of David.

From then, until the destruction of the Kingdom of Israel by the Assyrians and the exile of the ten tribes to Mesopotamia, a state of war existed between the Kingdoms of Judea and Israel. Only seldom did they make peace, in most cases to fight a common enemy. Sometimes, however, each joined forces or made an agreement with another enemy in order to harass his neighbor.

Jeroboam, after becoming King of Israel, was afraid that the continuation of the religious hegemony of the temple in Jerusalem might bring the people to turn back to the House of David. He knew also the religious and spiritual undercurrents which survived side by side with the worship of the God of the Hebrews. He thus made two calves of gold, and put one at Bethel (house of god) on the border with Judea and the other one at Dan in the north . The northern site is at the city situated close to the main spring from which the Jordan issues. This city was once Canaanite by the name of Laish, and was conquered by people of the tribe of Dan, who apparently did not like the semi-arid hills southwest of Jerusalem which were allotted to them. They travelled northward, on their way kidnapping a priest with the idol he was serving, suddenly attacked the Canaanites of Laish, burnt the town, and rebuilt it under the name of Dan. The kidnapped priest and idol became a cult center "until the day of the captivity of the land" (Judges 18:30). It seems probable that at least Laish-Dan was a Canaanite cult center connected with the big spring near by. The other source of the Jordan was later in the Hellenistic period called after the god Pan, from which came its present name Banias. It is hard to believe that such a big spring as the Dan could have remained without a God-Father. The worship of the golden calf apparently also persisted. Thus putting these two together, Jeroboam was able to reduce the cultural ties between the two kingdoms. The abstention of the people of the northern kingdom from the worship of the God in Jerusalem as their only god slowly brought about their adherence to the gods of the local Canaanites, especially the worship of Baal as the major deity. This had been adopted as the main cult when King Ahab married Jezebel, the daughter of the Phoenician king. In opposition to her and to this cult stood Elijah, the worshipper of the God of the Hebrews.

Baal was the one who decided whether the year would be dry or wet, whether there would be an abundance of food or hunger. Thus when a spell of dry years came on the land, Elijah challenged the priests of Baal to pray to their god and make him bring rain. When they failed, Elijah prayed to his god and when his prayers were answered he seized his advantage and slaughtered the priests of Baal. These deeds and many others did not deter the people of Israel from following Baal, the result being that when they were driven into exile they were not equipped with a special culture to differentiate them from the people among whom they settled, and so they gradually became assimilated.

The kings of Judea reigning in Jerusalem were officially expected to follow the religion of the House of David, a difficult task in a region whose people believed in Baal, Ashtoreth, and the other Canaanite deities. This was especially so when a period of truce was concluded between the two kingdoms and the influence of the northern religion on the southern people became possible.

During one of these truces, Jehoram, the King of Israel, asked Jehoshaphat, the King of Judea, to join him in an attack on Mesha, King of Moab, who had rebelled against him. They decided to take the southern road through the wilderness of Edom and attack Moab from the flank from where the Moabites were not expecting an attack, namely, from the desert. It seems, however, that the three kings (the King of Edom also joined the party) were good in tactics but not so good in army logistics. "So the King of Israel went, and the King of Judah, and the King of Edom, and they made a circuit of seven days: and there was no water for the host, and for the cattle that followed them. And the King of Israel said, Alas! that the Lord hath called these three kings together to deliver them into the hand of Moab!" (Kings II 3:9-10). King Jehoshaphat did not lose faith; he called for a prophet of Jehovah by the name of Elisha, the son of Shafat, "which poured water on the hands of Elijah". The prophet asks for music and when the minstrel played "the hand of the Lord came upon him, and he said: Make this valley full of ditches. For thus saith the Lord, ye shall not see wind, neither shall ye see rain, yet that valley shall be filled with water that ye may drink, both ye, and your cattle, and your beast ... And it came to pass in the morning, when the meat offering was offered that, behold, there came water by the way of Edom and the country was filled with water" (Kings II 3:15-17,21). The early sun shining red on the water made the Moabites believe that it was the blood of the Israelites and Judeans who had gone back to their old practice of killing each other, so they came down to collect the spoil, but were surprised by the Israelites and were defeated. The Bible tells only of the part of the battle which the Israelites won. It is quite enigmatic about the end of the war; it tells that the King of Moab took his eldest son, who should have reigned after him and sacrificed him upon the wall. "And there was great indignation against Israel; and they departed from him and returned to their own land" (Kings II 3:27). Another version of this war was found in Moab in a stele erected by the same King Mesha, in which he claims victory. It is not the task of this author to decide which of the ancient chronicles is right, though the vague way the Bible finishes the story of this war hints that the Moabite version is closer to the real facts.

On the other hand, the story of the "miracle" in the desert seems quite authentic. Flooding in the desert valleys of the Negev and Edom without any sign of local rain storm may happen from time to time, the author being a witness of such an event a couple of times. It seems quite probable that the armies got stuck in the Arava Rift Valley on their way to the southern flanks of Moab. This broad valley on the border of Edom, Moab, and Judea is extremely arid (less than 50 mm annual average precipitation) yet from time to time it is flooded by water flowing from the mountains of the Negev or Edom. The reflection of the red granitic mountains of Edom (Adom in Hebrew means "red") or the reddish color of the dawn above these mountains is also something which could be fact. It remains to the faith of the reader to decide whether the other parts of the story telling about the fate of the ancient people were also factual.

This story about a coalition between the kings of Judea and Israel to fight a foreign king reflects the threats which these countries faced during their existence. The two main enemies were the Egyptians in the south and the Arameans in the north. The Philistines, Moaabites, Ammonites, and the Edomites taking a share from time to time. The Assyrians and Babylonians entering the stage later.

In the fifth year of the reign of Rehoboam the son of Solomon, the enemy from the south Pharaoh Sheshak struck. The layers of charcoal and destruction can be traced in the many fortified cities he conquered [1]. Among these were Beer-Sheva, Arad, Gezer, and Megiddo. The ruined places were slowly rebuilt by the Kings of Judea and Israel. Those along the southern border were fortified and provided with a water supply system. At the Israeli citadel of Arad a tunnel was built, starting from the citadel and leading to cisterns dug into the chalky bedrock of Eocene age building the hill on which the citadel stands [1]. No connection was found between this tunnel on the citadel hill and the well of Arad described in the previous chapter, which is situated in the valley. It seems probable that the tunnel led only to the cisterns which were built to collect rain water. In times of emergency, either drought or the approach of an enemy, the cisterns could be filled up by hauling water from the well.

At Beer-Sheva, another interesting water system is found. The ancient city was surrounded by a thick wall of adobe bricks laid on a foundation of stones . In the eastern part inside the fortified city (Fig. 6.3) a deep shaft with spiraling steps leads, apparently, to the groundwater table of wadi Beer Sheva. A deep well to the depth of some 40 meters is located outside the walls near the main gate. An interesting additional feature connected with water is the canal, which drains the main street and passes below the main gate to the outside of the wall. It reaches near the deep well. Was it an artificial recharge project?

In the two cities of Arad and Beer-Sheva, altars were found. In Arad the entire, local sanctuary was also found. No idols were found in either place. It seems probable that the God of the Hebrews was worshipped there. These were the local cult centers against which the Bible spoke frequently, "Nevertheless the high places were not taken away; for the people offered and burnt incense yet in the high places" (Kings I 22:44). Although the main temple was in Jerusalem it appears that the people felt more secure with a local place of worship in addition to a dependable water supply system.

From time to time a pious king of the House of David would arise and call for a religious reform, drive away the priests of Baal and Ashtoreth, burn the trees dedicated to the Asherah, and call his people to the obedience of the God of Israel; but his descendants would again return to the worship of the gods of the Canaanite cults. One of the kings devoted to the god of his forefathers was King Uzziah (or Azariah). His interest was not only in the cultural and religious reform of his people but also for materialistic ends. One of his projects was the strengthening of Jerusalem by building towers to make the city more secure against its enemies. He was interested in the development of the arid part of his country, and in order to open this zone for settlement and provide water for his garrisons, he excavated cisterns in the desert near the small fortresses he had built. " Also he built towers in the desert and dug many cisterns: for he had much cattle in the low country and in the plains" (Chronicles II 26:10). Towers or citadels, having walls with inbuilt chambers, are found throughout the Negev and eastern Sinai, the chambered type of the walls, the system of cisterns and even the type of ceramics found in the ruins is typical, their time ranging all through that of the First Temple (ca. 970-580 B.C.). It seems probable that this region had its authentic population, most probably semi-nomadic, either from the tribe of Simeon, who dwelt in the southern semi-arid part of Judea, or Edomites. The towers are spread throughout the desert and not necessarily along the main roads. This may point to the fact that they meant to serve as guard-stations to protect the entire region against the nomads of the deserts, not only the caravan roads. The spread, and the fact that the towers and cisterns in the desert are connected, in the case of King Uzzia, with husbandry of flocks points, in the opinion of the author, to the fact that at the time of the First Temple the climate was more humid and the desert was more habitable. This can be seen, as a matter of fact, on the climate master curve (Fig. 1.3), where one can see that approximately between the years 1000-500 B.C. (3000-2500 B.P.) a depletion in the^{18}O and ^{13}C isotopes, a clear sign of a colder and apparently more humid spell.

The technical aspects of the cisterns of this period are also characteristic. They are located mostly on an impermeable marly layer on the hill slope below the tower, a collecting channel surrounds the hill slope and leads the water flow downslope into the uncovered cistern. There are also cisterns not connected to a tower; these were most probably built for the supply of water to a ranch located in small ravines or wadi beds, and get their water from the wadis. Pits serving as sediment traps are located at the inlet, some with collecting channels running along the slopes of the surrounding hills. A few cisterns may be grouped together to furnish an ample supply of water.

A water supply project, from the period of the First Temple but of an altogether different character, is that of the Niqba of the Siloam, which is the tunnel connecting the spring of Gihon and that of Siloam. The Canaanite part of this water supply project has already been discussed when the story of the conquest of the Jebusite citadel of Jerusalem by King David was told. Now the project of King Hezekiah will be discussed. According to the Bible (citations at the beginning of this chapter), the Niqba was dug in order to ensure a water supply during times of siege. For this purpose, he connected the spring of Gihon which was outside the walls of the City of Jerusalem, to the spring of Siloam, which was in the immediate vicinity of the walls.(Fig. 11.3).

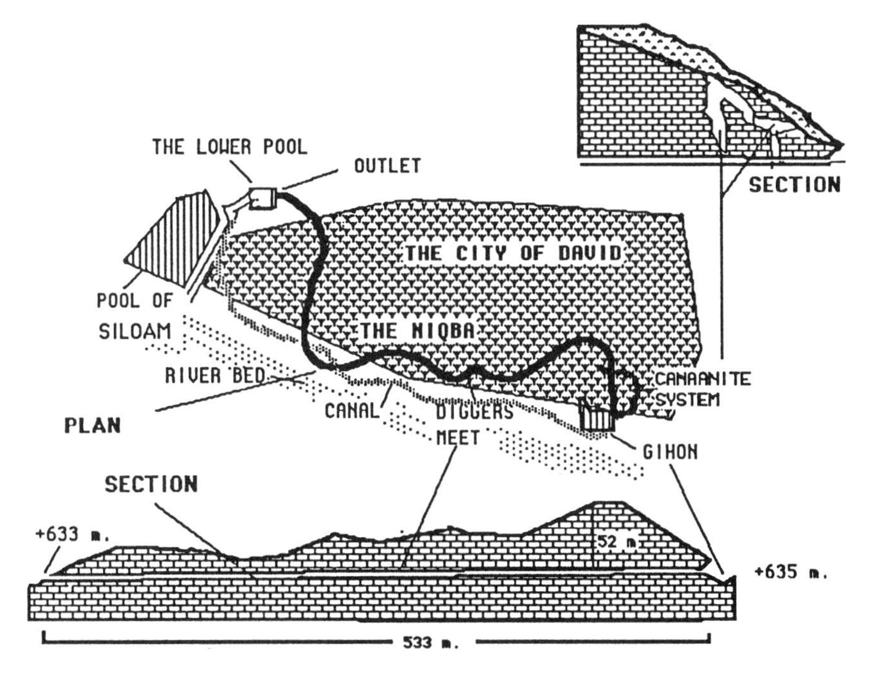

Fig. 11.3. Plan and section of the 'Niqba' in Jerusalem

Another document dealing with this project was found as an engraved inscription on the wall of the tunnel, and written in Hebrew letters characteristic of the period of the First Temple (Fig. 11.4). The inscription, the translation of which appears at the beginning of the present chapter, tells how the two crews of diggers working from each end of the tunnel met axe to axe after hearing each other for some time. The length of the tunnel is 533 meters, the difference in altitude between the two interconnected springs is 2.1 meters and the height 1.1 to 3.4 meters. The distance between the two interconnected springs is only 320 meters. This means that the tunnel makes a rather long detour (Fig 11.3). This was an enigma to the archeologists who investigated this project; one of them suggested that the wide detour was made in order to avoid passing the water under the tombs of the kings of the house of David, due to the Jewish tradition that the remains of the dead defile everything, even flowing water [2]. This enigma was solved after the hydrogeological investigation showed that the system of springs issuing from the limestone of Turonian (Cretaceous) age is a karstic system, including the pulsating spring of Gihon. The tunnel seems now to be following a karstic channel which was connecting the two systems.

This also answers the question of the diggers meeting, although not following a straight line, as well as the unnecessary height of the tunnel. Also the story told in the inscription of the diggers hearing each other before meeting is better understood

[3,4,5,6]. The excavations carried out by the late archeologist Yigal Shiloh in the City of David confirmed the karstic explanation.

Fig. 11.4. The 'Niqba' inscription and facsimile, now in the Archeological Museum, Istanbul, Turkey

The Niqba inscription (now in the Archeological Museum of Istanbul) contains also an enigmatic Hebrew word which is "Zade". In the opinion of the author the subterranean karstic solution channel solves this enigma. The word in his opinion, means "way" or "direction" as the root of the word is "z-d-h" from which comes the word "mezid" meaning "to do something on purpose".

Upon viewing the whole ancient water supply system of Jerusalem (Fig. 11.3), the evolution can be said to be the following: The first stage was the digging of a horizontal tunnel to a small spring just outside the town. In the second stage, this tunnel was joined by another declining tunnel leading to the inside of the walls. When this spring dried up, two pits to look for the groundwater table of the Gihon were excavated, most probably along ancient karstic solution channels. This failing, the upper tunnel was connected by another tunnel to the spring of Gihon. As this spring was too far away from the walls, and as an existing narrow karstic solution channel showed that the two springs were connected, two teams of diggers were ordered to start digging from either end to try to complete the water project as quickly as possible to help Jerusalem withstand a siege by the approaching enemy.

Thus, when the Assyrian King Sennacherib returned to punish King Hezekiah and the people of Jerusalem for trying to remove the yoke of Assyria by going into a coalition with the King of Egypt, Jerusalem would not surrender. Sennacherib sends "Tartan and Rab-Saris and Rab-Shakeh from Lachish to King Hezekiah with a great host against Jerusalem" (Kings II 18:17). The Assyrian army starts the siege by taking control of the water supply to besieged Jerusalem "And when they were come up, they came and stood by the conduit of the Upper Pool which is in the highway of the Fullers Field" (Kings II 18:17). The Assyrians were informed about the coalition with Pharaoh yet they were not aware of the new water system and were therefore sure that Jerusalem would surrender.

"And Rab-Shakeh said unto them, Speak ye now to Hezekiah ... Now behold thou trustest upon the staff of this bruised reed even upon Egypt, on which if a man lean it will go unto his hand and pierce it ..." (Kings II 18:19-21).

And when the counselors of Hezekiah, standing on the wall, asked him to speak Aramaic and not Hebrew, the Assyrian answers, "Hath my master sent me to thy master, and to thee, to speak these words? Hath he not sent me to the men which sit on the wall, that they may eat their own dung, and drink their own piss with you?" (Kings II 18:27).

And Rab-Shakeh goes on speaking Hebrew to the people on the wall promising them that if they betray their king and surrender the city to him, he will not treat them in the cruel way the Assyrians were known to treat rebels, but: "I come and take you away to a land like your own land, a land of corn and wine, a land of bread and vineyards, a land of olive oil and of honey" (Kings II 18:32).

The prophet Isaiah, the son of Amoz, strengthens the spirit of the king and his people by promising that the salvation of Jerusalem is yet to come, and indeed a plague strikes the Assyrian army. Sennacherib retreats and is later murdered by his sons. The small kingdom of Jerusalem gains a period of 95 years to survive. This century, from the time of Hezekiah to that of Zedekiah, the last King of Judea, was a crucial period from the cultural point of view and for the preparation of the spirit of the people of Israel for survival as a nation even in exile. During this period, two decisive and even extreme religious reforms were introduced. The first by King Hezekiah (Chronicles II 29-31), the second by King Josiah, a great-grandson of Hezekiah. A copy of the Book of Laws found in the Temple during its reconstruction gave the King Josiah the opportunity to wipe out all the Canaanite religious rites which had been assimilated into the Jewish religion and were even practiced in the Temple of Jerusalem (Kings II 23; Chronicles II 34,35). He also destroyed all the local altars where God was worshipped, like those in Beer Sheva and Arad, mentioned above. These reforms uprooted the Semitic-pagan cults which had become part of the religious rites of the Jews, thus fortifying them against assimilation into the religion of Babylon to where they were exiled when Jerusalem was captured in 587 B.C. Without these reforms and without the socio-moral message which was the main theme in the prophecies of Isaiah and Jeremiah, the fate of the people of Judea would have been the same as that of the other tribes of Israel exiled by the Assyrians to Mesopotamia where they became assimilated into the surrounding Semitic-pagan culture and vanished from the stage of history. Was the "Niqba project" a material tool which enabled the religious, cultural, and national survival of the Jewish people?

As will be related later, King Sennacherib was also a master in waterworks. He built canals and dams far more magnificent than those of King Hezekiah of Jerusalem, yet when Assyria was conquered by the Babylonians, this colossal empire vanished too, and only the rich archives of cuneiform tablets, buried under the ruins of its towns, were left to tell the story of these people.

The moral teachings of Isaiah and Jeremiah, adopted by the Jewish people, survived long after the Niqba and all the water systems of Jerusalem were abandoned and forgotten. In the contest between material and spiritual values, the latter seem to have a better chance to survive. It might be that this was foreseen by Isaiah when he prophesied "Ye have seen also the breaches of the City of David that they are many and ye gathered together the water of the lower pool. Ye made also a reservoir between the two walls for the water of the old pool, but ye have not looked unto the maker thereof neither had respect unto him that fashioned it long ago" (Isaiah 22: 9-11).

Going back to King Josiah, his piety was of little help to him when he refused to allow Pharaoh Necho (610-595 B.C.) to pass through his country to fight the Babylonians. He was killed in the battle in the Valley of Megiddo, to be mourned by all the people of Jerusalem and the prophet Jeremiah (Chronicles II:35). The same Pharaoh Necho is famous for one of the greatest pioneering hydraulic works, namely, a canal from the Red Sea to the Nile forming a connection between the Indian Ocean and the Mediterranean, a project which was rejuvinated by Darius, the King of Persia, when Egypt became part of the Persian Empire.

The first emergence of the Assyrians as an independent nation was at about 2000 B.C. when the Sumerian kingdom of Ur collapsed before the invading tribes from the east and west. As already explained, the author's conjecture is that the collapse of Sumer was connected with a spell of aridity starting around 2000 B.C., which caused the ingression of the western Semites or Amorites from the deserts of the Middle East into Mesopotamia. At about 1700 B.C. there was another wave of migration, this time from the north. These were mainly Indo-Aryan tribes dwelling on the steppes of Central Asia, who were able, with the use of horse- drawn war chariots, to overcome their opponents and penetrate into the fertile lands to the south. They invaded Persia and also overwhelmed the civilizations of the Indus Valley and became the dominant race over India. Another group of tribes, the Kassites, came down into the Euphrates Valley and settled on the border of Babylonia. There is no direct evidence that this wave, which continued through the first part of the second millennium B.C., was a result of a climatic change, yet the second wave of Indo-Aryan tribes at about 1400 B.C. was apparently connected with a cold spell . The disappearance of the late Minoan culture in Crete was connected with a series of invasions from the north by Indo-Aryan Achaean tribes, who pushed the former Indo-Aryan settlers of the peninsulas and islands of Greece and Anatolia southward.

This shift of nations reached its climax in about 1200 B.C. when a new wave of invading tribes caused a shock wave to spread over the entire Middle East. In the history of the ancient world this series of events is connected with the fall of Troy to the Achaeans, which symbolized the fall of the Mycenaean culture (the inheritors of the Minoan culture) and the beginning of the Greek period "proper". The onslaught of the Achaeans led to the attacks by the "People of the Sea" on the southern and eastern shores of the Mediterranean (events which have already been described in former chapters) and also led to the settlement by the Philistines, Sicilians, Tyrrhenians, Etruscans, and Sardinians all along the shores of the Mediterranean. Repercussions of these shock waves also penetrated the continent. By a series of events resembling the collapse of a row of dominoes, the onslaught of the People of the Sea caused the fall of the Hurrian kingdom of Cilicia. The population of this kingdom pressed inland from the sea, fled into Anatolia and caused the destruction of the Hittite Empire which, together with Egypt, was the main power to balance that of the kings of Mesopotamia. At the place where the Babylonian-Hittite-Egyptian empires of the Bronze Age sporadically contested the Fertile Crescent, the kingdoms of Aram and Israel were formed.

A new power, the Assyrians, was soon to replace them all. The Assyrians, a branch of the Akkadians, settled on the upper stretches of the Tigris when the latter settled in lower Mesopotamia, around 3000 B.C. Their culture also depended on river irrigation, although the country was less arid than the south. Their religion was very similar to that of their neighbors to the south. In many respects, Babylon was regarded by the Assyrians as a holy city. During most of the second and third millennia B.C. the Assyrians played a minor role in the fate of the Middle East, as they were restrained either by the Indo-Aryans, who ruled the kingdoms of the Mitannites and Hittites to the north and northwest, of the Akkadian-Semitic kingdom to the south. In about 1200 B.C., an Assyrian king conquered Babylon and took the statue of Marduk into captivity. Yet the Assyrians, mainly because of religious respect, did not continue their pressure southward towards the fertile lands of lower Mesopotamia. Their eyes were turned toward the west, from where they were harassed from time to time by the western semitic peoples, the Arameans and Amorites.

These peoples of the mountains of Lebanon and Syria were a constant threat to the people of the valley of Mesopotamia so the Assyrians developed a strategy of raids into these countries as a means of defence and later as military expeditions to plunder and collect annual tribute. The mountainous nature of the land between northern Mesopotamia and the Mediterranean helped divide the inhabitants. Thus the Assyrians adopted a policy of "divide et impera", promoting and helping countries who were ready to pay them tribute against those who refused.

Two important war instruments aided the Assyrians in this policy. The first was the cavalry assisting the war chariot which was adopted from their northern neighbors, and the second was the siege machine. The horse was better adapted to the Assyrian plain than to those of Babylonia and Sumer, as it was nearer the highlands and steppes of Anatolia, and enjoyed more rain. The siege machines were a new development which, backed by the sound logistics of supplying a tremendous army, enabled the Assyrians to sweep the Middle East. Following the conquests, the local people were deported to other provinces and other people settled in their place. Thus, after Sargon II (721-705 B.C.) conquered Sumeria he took the local Israelites into exile and settled them on the Khavor river, replacing them with a foreign people who acquired the name of Shomronites or Samaritans.

Part of the story of his son, Sennacherib (704-681 B.C.), who warred against Hezekiah, King of Judea, has already been told. The same Sennacherib constructed one of the biggest water projects of the ancient world. As he did not like the site of the old capital of Assyria, Dur-Sharrukin, which lacked gardens and parks, he built a new one, Nineveh, on the confluence of the Tigris and the Khosr rivers. He surrounded the new palace which he built there and the walls of the city with gardens. To irrigate these gardens by gravity he constructed a system of canals about 220 km long, which reached the foothills region. This system crossed the river-beds on its way by means of raised aqueducts. One of them, at the present village of Jerwan, is about 300 meters long, 20 meters wide and 7 meters high, [6]. At the source of this system in the hills of Armenia is an inscription by Sennacherib which is most probably a transcript of the king's ceremonial speech for the inauguration of the canal:

"To open the canal I sent an Ashipu priest and a Kalu priest ... and I presented gifts to Enbilulu Lord of Rivers and to Enemibal (Lord who digs canals), The sluice gate like a flail was forced inward and let in the waters in abundance. By the work of the engineer its gate had not been opened when the gods caused the water to dig a hole there in... At the mouth of the canal which I had dug through the midst of Mouth Tas I fashioned six great stele with the images of the great gods and my royal image in a gesture of obedience. Every deed of my hands I wrought for the good of Nineveh I had engraved thereon. To the kings my sons I left it for the future" [6]. Sennacherib imported from Egypt the well-sweep device (shadoof), to enable his people to pump water from the canals. The introduction of this innovation was carved on the walls of his palace.

The budget for the construction of gigantic water supply systems and their maintenance undoubtedly came from the tribute and booty collected by the Assyrian war machine, part of which came from King Hezekiah who decided to surrender. The atrocities perpetrated by the Assyrian army caused many kings to agree to put themselves at its mercy and pay the heavy tribute to spare their country from total destruction and their people from exile.

The son of Sennacherib, Esarhaddon, decided to ensure the southern border of his empire against the intriguing Pharaoh. He invaded Egypt (671 B.C.), destroyed its capital, Thebes, and carried away two granite obelisks to Nineveh. He ruled Egypt for 15 years. In his annals describing this campaign he tells about the problems of ensuring an adequate water supply for his troops. He complains that though the area was called The Brook of Egypt,(probably Wadi El-Arish) there is no river along the route and he had to draw water by buckets from wells [7]. The wells of El Arish were to be used for many more armies in generations to come crossing the Sinai Desert westward or eastward.

At this point, Assyria reached its peak of power. The time for the back swing of the pendulum had arrived. The people of Babylon, being unable to live under the heavy yoke, revolted. Their king, Nabopolassar, made a coalition with the Medes who were living on the Iranian plateau, free from Assyrian oppression. They challenged the army of the Assur and, after a fierce battle, defeated it, and sacked and burnt Nineveh and the other big cities (612 B.C.). This was the end of the Empire of Assyria. Unable to maintain the huge public water supply system and its urban society due to the lack of income from tribute and booty, the country was quickly impoverished and returned to being a rural society. It became a neglected province of the Babylonian and, later, Persian empires. The son of Nabopolassar, Nebuchadnezzar, turned the capital Babylon into one of the wonders of the ancient world by constructing a sophisticated water supply, presumably by syphon pipes, which enabled water to be lifted to the roofs of the buildings and gardens to be planted there.

In order to ensure the southern extension of the empire he inherited from the Assyrians, he sent his army against Zedekiah, the King of Jerusalem, who rebelled and concluded an alliance with the King of Egypt. This time Jerusalem did not withstand the siege, which lasted about 2 years. The Bible tells us that there was no food in the city due to the siege. Nothing is said about the water supply, which might have been functioning throughout the siege. When the walls of Jerusalem were breached by the Kassites and Babylonians, Zedekiah tried to escape through the gate of the wall facing toward the Judean Desert. He was caught near Jericho, his sons were killed before his eyes and he was blinded. The aristocracy of Jerusalem was then taken in exile to Babylonia. There, on the canals of the Euphrates, called in Akkadian "naharot" or rivers, they lamented the destruction of their temple.

"By the rivers of Babylon there we sat down, yea, we wept when we remembered Zion. We hanged our harps upon the willows in the midst thereof" (Psalms137:1).

A successor of Nebuchadnezzar, Nabonaid, tried to establish a cult of the Moon-God, Sin, as the main god instead of Bell-Marduk and to expand his empire into the deserts of the south. He invaded Arabia and established a center for the worship of the moon at the oasis of Teima. While he was staying there, the priesthood of Bell-Marduk rebelled against his son Belshazzar, and opened the gates of the city to Cyrus, the Persian, an obedient worshipper of Ahuramzda, the God of Righteousness and Light.

One will recall that the Persians and Medes who followed Cyrus the Great in conquering the Babylonian Empire that had vanquished the Assyrian Empire were of Indo-Aryan origin. Their language remained, and even today is rather similar to Sanskrit and thus to modern Hindi. One may still find in the syntax and words many similarities to the Latino and Germanic languages. Thus, while water will be "ab", similar to "eau" and "aqua" ("panj-ab" means five waters or rivers), in Iran to this day, God is still pronounced "Ghoda", and the old gods or false gods in the inscription of Xerxes, where he describes his devotion to Ahuramzda, are "daiva", reminding us of "deus", "dieu" and "divine".

The Indo-Aryan people invaded the Iranian plateau in several waves, as similar tribes had done in Greece and India. The most prominent was that at the beginning to mid-second millennia, between 1700 to 1400 B.C., and the last occurred in the first millennium B.C. The invaders were a few tribal groups who spread over the plateau, subduing and absorbing its former "Caucasian" population. The group of tribes occupying the southwestern part of the plateau were called Persians, while those who occupied the northwestern part were called Medes. Their capital city was Ecbatana (or Hamadan of present times). They are first mentioned in 836 B.C. by Shalmaneser III, King of Assyria, as the enemies of his country. In 612 B.C. they joined the Babylonians and sacked Nineveh. The Medes were later subjugated by the Persians who annexed their land to form the "Land of Persia and Medes" ruled by the Persian royal house that traced back to an ancestor called Achaemenes, dwelling at Pasargadao in central Fars. Their main capital was at Susa (Shushan of the Book of Esther) located on the eastern border of the Valley of Mesopotamia. The king who accomplished the formation of the Land of the Persians and Medes was Cyrus I, called "the Great." After his success in fortifying the borders of his country against the semi-nomadic tribes on the east, and conquering Anatolia, comprising Lydia and Ionia, he expanded his territory over the Babylonian Empire. After he took control of Babylon in 539 B.C., proclaiming himself its king, he proclaimed his permission to the Jews in exile to return to their land and rebuild their temple.

Cyrus believed in Ahuramzda, the Lord of Light and Goodness. Although paying homage to the local gods was part of a general policy for winning over the people of his vast empire, the Jews believed that there was a special concern toward their religion as at that time the Zoroastrian religion of Cyrus was also monotheistic.

There are many wonders beyond the comet-like rise to power of the Persian Empire. One is the fact that the Iranian plateau is a vast, harsh desert surrounded by lofty mountainous ranges which obstruct the inflow of rain-bearing winds either from the Mediterranean Sea to the west, the Caspian Sea to the north, or the Indian Ocean to the south. Yet the barriers themselves, namely the Zagros mountains on the east and the Alborz mountains on the north, which reach altitudes of three to four thousand meters, as well as the other lofty mountain ranges traversing the Iranian plateau, receive an ample supply of rain and snow from the incoming storms. The water flows down the mountain slopes in torrential floods and are partly absorbed in the vast alluvial fans built of gravel. These fans stretch along the foothills. The remaining water flows into the inland basins to form huge saline lakes and marshes. The ingenuity of the Iranians is demonstrated by their ability to utilize these floods. In the first place, they utilized the downward flow of the rivers from the mountains by terracing the mountain valleys and irrigating them by canals which branched off from the main river-bed. However, most of the rivers of Iran flow mainly during the winter and spring, while the summer and autumn are long, hot, and dry. For the purpose of obtaining water for irrigation also during the dry season and dry years, the Iranians developed one of the most ingenious hydrological devices, namely the "ghanat" or "qanat" (called "khariz" in the Turkish-speaking provinces and "foggara" in the Arabic-speaking provinces) or referred to today as "chain of wells".

The first mention of this system is by the Greek historian Polybius (204-122 B.C.) who, about a century after the conquest of Persia by Alexander the Great, visited Ecbatana the former capital of the Medes [2]. The hydrological principle of the ghanat is the excavation of a big subsurface drain into the gravel layers which cover the foothills of the mountain ranges surrounding and criss-crossing Iran, to drain the tremendous quantities of groundwater which infiltrated from the floods. To a man flying over the vast stretches of desert of Iran, thousands and thousands of mole-like holes are revealed, arranged in lines stretching from the foothills into the plain. A village can be seen at the lower end of the line with its fields and orchards irrigated by the water flowing from the subsurface. These mole-like holes are vertical shafts excavated at specific distances from each other, from the surface to the gallery in the subsurface. They were dug to ensure access to the surface, for removal of soil excavated from the gallery, and for ventilation. Later, the shafts were used for the maintenance of the galleries (Figs. 11.5,11.6).

The author, who for several years worked in Iran as a hydrologist, had the opportunity to investigate many of these ghanats and discuss with the "mughani", the people who are specially trained to dig these galleries, the various methods of location and excavation. This practice of excavation, as the reader can see, has continued from the time of the Achaemenid Empire (and possibly earlier) to the present day, namely, more than 2000 years. One might say that, in principal, there has been little change in either methods or tools (Fig. 11.7).

Since ancient times, the people of the Iranian deserts knew that a large portion of the floodwater flowing from the mountains was absorbed into the porous gravel layers which were deposited by the same streams. They also knew that the water in the subsurface flows very slowly, and finally emerges at the bottom of the plains, flowing as seepage into the lakes and marshes which, because of the high evaporation, become highly saline. Thus, they could not use the water where it seeped out naturally, as the soil in the lower stretches of the plains was very saline due to the upward movement of the water through the capillaries of the soil and the salinization of the soil. They had to locate the water in the upper parts of the gravel fan, but there the water was deep below the surface and they did not have the mechanical facilities to pump it up in big enough quantities to irrigate large enough areas. The solution they found was to drive a gallery into the subsurface with a gradient smaller than that of the land surface but greater than that of the groundwater table (Fig. 11.5). Thus gravity caused the water in the gallery to flow, draining the saturated subsurface layers.

The digging of a ghanat begins at the outlet end, thus guaranteeing outflow of the water so that the mughani could work in a drained gallery. The surveying and siting and planning of the ghanat is done by an experienced mughani. At the request of the owner of the land, he surveys the foothills area and drives a few exploration wells ("cha" in Farsi which is the modern Persian language) in which he measures the depth of the water tables. From there, with the aid of very simple surveying tools, he locates the outlet of the ghanat ("cheshmeh", i.e., "eye" and "spring" in Farsi) from where the excavation starts. At the same time, the line of the ghanat is marked on the surface and the diggers start to drive vertical shafts to meet the horizontal gallery. As the pace of digging of the horizontal gallery is twice that of the vertical shafts, the distance from the shaft to the outlet of the gallery and then to the neighboring shaft is twice the depth of the shaft. As digging proceeds, the quantity of water draining from the gallery increases. In order to keep the walls of the gallery from collapsing, their dimensions are kept to a minimum, just large enough for a mughani to creep inside. It is usually an oval-shaped hole of 90 x 40 cm. If the walls start to collapse, oval-shaped rings are prepared, usually of baked loam, to insert into the hole as work progresses. The elliptic form of the rings enables them to be inserted while tunneling continues without having to reduce the diameter. The direction inside the tunnel is kept in a straight line by a row of lamps left behind the workers which should look like one light all the time. The shafts enable them to remove the excavated sand and gravel without having to crawl the long distance to the outlet. The shafts also keep a draft of air flowing through the tunnel. If the gallery is very deep and there is an accumulation of gases, which the mughani can observe according to the flame of their oil lamps (today they use carbide lamps), they put sails up at the outlets of several shafts to catch the wind and drive it inside the gallery. The special mixture of gravel, sand and fine silt of the alluvial layers usually keeps the walls standing while digging is in progress, especially when the protective rings are inserted, but over the years the walls collapse and then the mughani will go in through the shafts to clean out or build a by-pass tunnel. Sometimes an accident may happen while the tunnel is being excavated. To the foreigner working in Iran, the lack of precautionary measures, as well as the fatalistic approach of the diggers, not to say the landowner, to such an accident and loss of life is shocking, especially in regions where the composition of the soil makes the likelihood of a collapse greater. In such regions, they have boys to help them carry the excavated material from the tunnel, thus its diameter can be kept at a minimum. In one area in central Iran, where there was a high incident of accidents, the mughani worked in their burial clothes, so that if the tunnel collapsed on them their comrades did not have to dig them out, and wash and dress them in order to give them a proper burial.

When there is a sufficient flow from the ghanat, excavation stops. It might also stop if the gallery reaches bedrock. The last shaft, "madar chah" (Mother-well), is the deepest one, from which excavations restart if the water of the gallery becomes depleted, as happens in a series of dry years or if the owner needs more water. If the gallery and the "madar chah" reach bedrock, the excavation may bifurcate to form a few branching galleries draining into one "cheshmeh".

Although the outflow from the ghanats fluctuates according to the seasonal change in the level of the groundwater table, which is lowest in autumn and highest in springtime, the ghanat system guarantees an ample supply of water to the farmers even in years when the supply from floodwaters fails due to drought. The length of such a ghanat may reach a few tens of kilometers and the depth of the "madar chah" more than 100 meters.

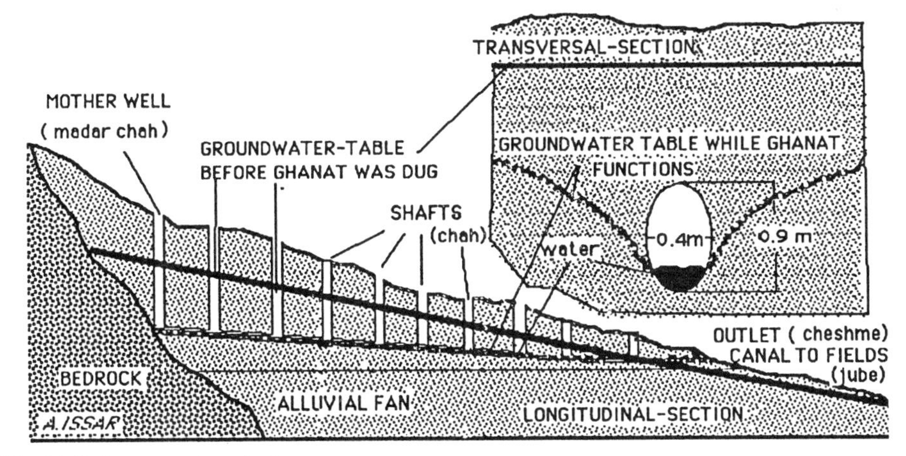

Fig. 11.5. Cross section of ghanat

The origin of the ghanats is, apparently, in the mountain valley river-beds where the people dug galleries to catch the subsurface water flow in the gravel beds. Later, they descended to the wide plains of the plateau and expanded the system all over the plains of the plateau of Iran. It can be claimed that the ghanat system enabled the settlement of the arid plateau by a sedentary agricultural society which could afford to maintain a high standard of living in urban centers surrounded by orchards. Rich households had their water pools supplied from the canal ("jube") bringing in ghanat water. The pool was surrounded by a garden ("firadus"), once called "pairidaeza", namely, roundly formed or planned, and clearly the root of the word "paradise".

A secure water supply system guaranteed that Iran changed from a desert with a few scattered oases to an agricultural society ruled by the former tribal chieftains who became feudal landlords governing tremendous stretches of land. These landlords maintained much the same way of life, and played the same sports, such as riding, archery, and hunting, as their tribal chieftain ancestors. Slowly, the system of government became centralized and Iran became fused into one kingdom.

In order to maintain control over the lands and submission to the central government, Iran developed a system of communication by horseman and fire signals, an echo of which can be found in the book of Esther: "And he wrote in the name of King Ahasuerus, and sealed it with the king's ring and sent letters by posts on horseback, and riders on mules, camels and young dromedaries." (Esther 8:10). When Cyrus started his campaigns after forging the Persians and Medes into one kingdom, he was backed by a rich, fertile country supplied by water from natural and artificial rivers, governed by a feudal system descended from the ancient Indo-Aryan warrior tribes excelling in horseback riding and archery, and a huge peasant society from which large hordes of men for his infantry could be raised. The vast country was criss-crossed by a system of roads along which government posts travelled from one station to another, delivering commands and bringing news.

This system was extended to the conquered provinces as the Empire expanded, reaching, at its peak, the border of Libya in the southwest and India in the southeast. ("This is Ahasuerus that reigned from India even unto Ethiopia over a hundred and seven and twenty provinces" Ester,1:1). The practical knowledge for excavation of the ghanat system spread with the empire. Thus, after Cambyses conquered Egypt in 525 B.C., the Persians constructed ghanats also at the oases of the Western Desert of Kharga and Dakhla [6]. Interestingly enough, this time the source of the water was not the gravel fans fed by flood water but the fractured Nubian sandstone from which fossil water flowed. The same happened in the Arava Valley, south of the Dead Sea, where a ghanat system was driven into the cliff to reach the Nubian sandstone and catch the flow of fossil water which, until then, flowed into a salt marsh.

In the mountains surrounding Jerusalem, the author has observed that in many of the natural springs flowing from the fractures of the limestone, galleries and shafts were excavated leading into a storage pool. The shafts were dug at fixed distances from each other, the entire system being almost identical to the spring-ghanat system observed by the author in the mountains of Iran. It is thus quite probable that the remnants of the ghanats found in the Jordan and Arava Valley's and the galleries and shafts found in the Judean mountains were excavated during the time of the Persian dominion. Although the spring-galleries might have been older, their reconstruction by adding a shaft every few tens of meters undoubtedly bears the imprint of a Persian ghanat constructor or "mughani".

After Iran was conquered by the Arabs, the ghanat system spread with the Moslem conquests. One finds ghanats also in Spain. Madrid used to obtain its supply mainly from a system of ghanats, (Prof. Ramon Llamas of The University of Complutense claims that the name Madrid comes from the term "madar chah"-mother well). The Persian water system, however, was not based solely on ghanats. As the kingdom of Persia extended into the province of Elam in eastern Mesopotamia, the Persians learnt the methods of canal construction and practiced them to supply their cities and estates in the lowlands. Also in the high plateau they made use of this technique to divert the flow of floods and rivers to their fields and orchards. The contemporary traveller in the arid plateau of Iran may frequently witness the reconstruction of a damaged canal or the excavation of a new one by a group of peasants. Sometimes the gradient of the canal is so small that to the visitor it looks as if it goes upslope. At this request, one of the peasants may volunteer to show him the homemade traditional geodetic survey instruments with which they calculate the slope, and the traveller may be sure that, when finished, the canal will transmit the water to where it was planned.

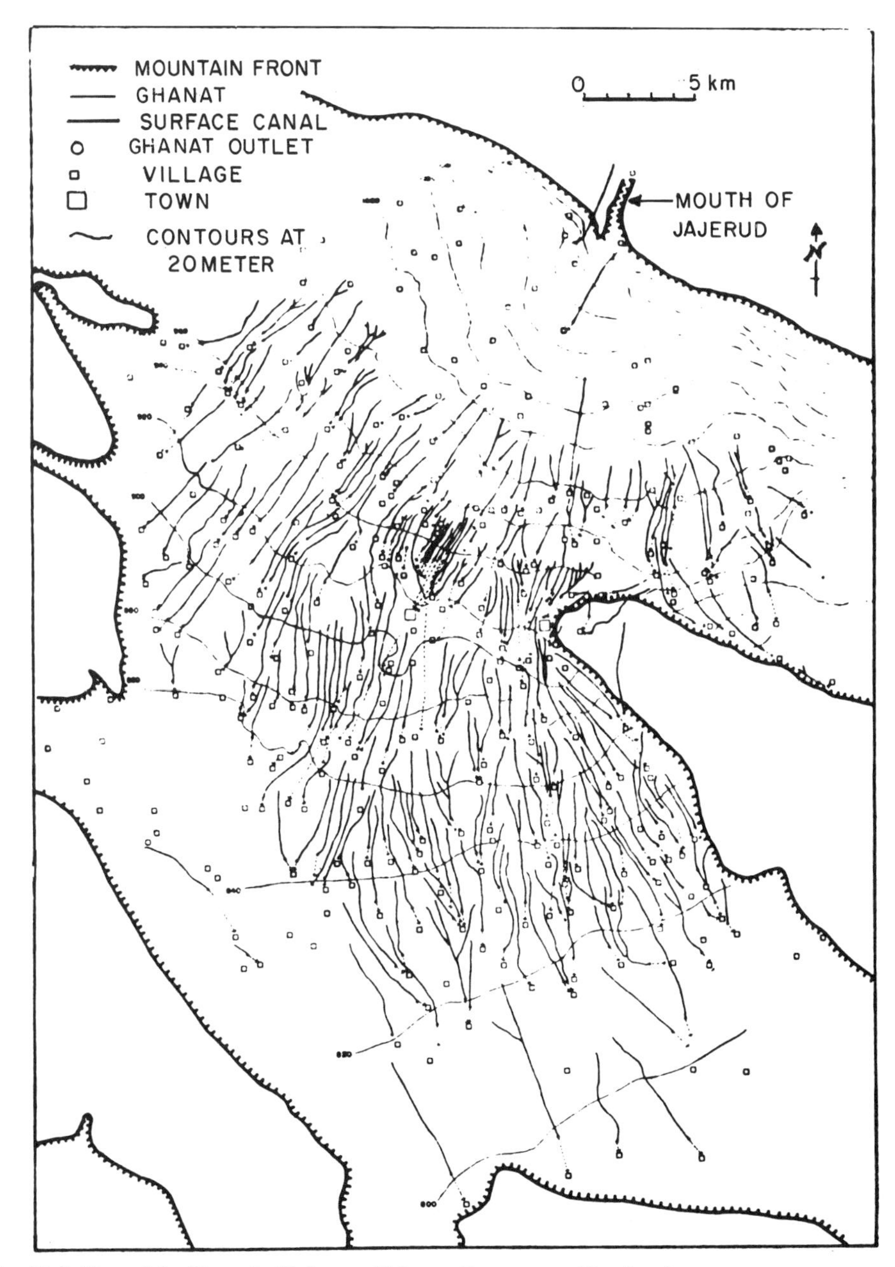

Fig. 11.6. Map of the Varamin Plain near Teheran, Iran, crossed by the ghanat system

Fig. 11.7. Mughani-ghanat diggers and the author, Iran 1964

The skill, industriousness, and cooperation of the canal diggers advanced during the generations to such a level that a European waterworks construction engineer told the author that in many cases he found that the digging of canals by local hand labor competes in efficiency and cost with construction by machinery. It might be that the skillfulness of his engineers and canal diggers made Darius I decide to rejuvenate one of the biggest projects of his time, namely, the connection of the Nile river with the Gulf of Suez, which had previously been done by the Egyptian Pharaoh Necho, but which had silted up and fallen into disuse. Traces of this canal can still be seen on the ground in the Eastern Desert of Egypt and on air photographs (Fig. 11.8).

Darius commemorated the completion of this project in the following inscription [9]:

"A great god is Ahuramzda, who created yonder sky, who this earth created, who created man who created welfare for man who made Darius king, who bestowed upon Darius the King the kingdom, greatly rich in horses, rich in men. I am Darius the Great King, King of Kings,... saith Darius the King, I am a Persian. From Persia I seized Egypt. I ordered this canal to dig from the river by name Nile which flows in Egypt to the sea which goes from Persia. Afterward this canal was dug thus as I commanded, and ships went from Egypt through this canal to Persia, thus as was my desire."

As can be seen from this inscription, Darius referred to his god Ahuramzda as his benefactor without mentioning other names, and indeed from many other inscriptions it is obvious that at the time of the first Achaemenid kings the teachings of Zoroaster were adopted as the state religion. It also seems quite possible (although it has not yet been proved) that Zoroaster himself lived and taught during the time of Cyrus the Great and succeeded in converting Darius' father, Hystapses, who was a King of Persia under Cyrus, the king of kings (559-529 B.C.). Thus after Darius I (the Great) won the kingdom following a rebellion after the death of Cambyses II, he proclaimed that his success was due to Ahuramzda, who came to his aid. This proclamation was inscribed in cuneiform writing on a cliff at Behistun, east of Ecbatana (Hamadan) in three languages, Persian, Elamite and Babylonian (Akkadian). The deciphering of this inscription by Sir Henry Cheswicke Rawlinson was the starting point for deciphering all the cuneiform scripts and thus all the writings in Akkadian and Sumerian. For the story of its deciphering it is suggested that the reader refers to Ernst Doblhofer's book Voices in Stone [10].

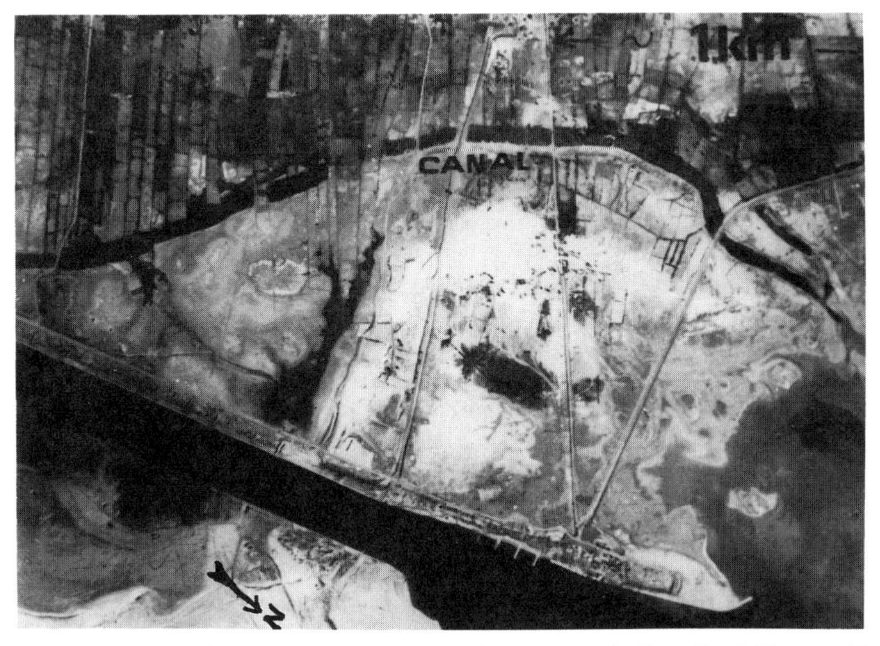

Fig. 11.8. Traces of Darius' canal as seen on an air-photo, west to the Suez Canal (Courtesy T. Weissbrod, Geological Survey, Israel)

Zoroaster's religion was based on the ancient Aryan heritage in which the great powers of nature were worshipped. The same heritage was brought into India and formed the basis of Hinduism with Brahman as a relatively monotheistic entity governing a vast assemblage of minor deities. These were divided into two groups, the good ones "devas" and the bad ones "asuras". The Iranian Aryans did just the opposite--the "devas" or "daivas" became the demonic divine powers while the "asuras" or "ahurez" became the positive powers. Zoroaster singled out Ahuramzda, the God Who Knows All (or God of Full and Right Knowledge). Although Ahuramzda was the supreme god, other deities, sometimes presented as qualities of Ahuramzda, sometimes as demi-gods, were functioning in the World of Light, e.g., the God of Righteousness, of Good Thought, Holy Piety, and even Wish for Dominion. Against the good world stood the world of evil and lies. There was a constant contest between Angra-Mainyu (Ahriman) or Evil Spirit and Spenta-Mainyu or Holy Spirit. According to this faith, man can choose his place whether on the side of Ahuramzda and his aids or on the side of the demonic daivas of evil and lies. The end of the contest is at the "end of days", when Ahuramzda will triumph. A flood of molten metal will flow over the world and all evil will be burnt up and the good left unharmed.

It seems, however, that while the kings adopted Zarathustra's (Zoroaster) monotheism, the polytheistic religion persisted in Persia. Thus Xerxes, the son of Darius (486-465 B.C.), proclaimed in an inscription discovered in Persepolis that "by Ahuramzda's will he destroyed the temples of the daivas and ordained that daivas shall not be worshipped". Yet the worship of other deities continued and in the inscription of Artaxerxes II (404-359 B.C.) one finds mention of the new old gods, Mithra and Anahita. "Says Artaxerxes, great king, king of kings, king of lands, king of this earth: the hall of columns (Apadana) Darius I my great-grandfather had constructed. Later on, during the reign of my grandfather Artaxerxes I, had burnt down. By the grace of Ahuramzda and Anahita, and Mithra, I reconstructed this Apadana. May Ahuramzda, Anahita, and Mithra protect me from all injury and may they neither injure or destroy this (Apadana) which I have constructed".

The combined power of the three deities did not help against the fire which burnt the original scriptures of Zoroaster and which is said to have been set by the drunken soldiers of Alexander of Macedonia about 20 years later. Yet before the final destruction of this tremendous empire, the degeneration of the ruling class was obvious. This was expressed not only in recurring court conspiracies, upheavals, murders, and poisonings but also by the regeneration of the primitive deities.

Mithra, the Aryan sky god, mentioned by Artaxerxes, was also worshipped as the sun god and the god of light, also in India. Anahita was the goddess of the rivers and brought fertility. She was a spring-fed river which poured down from the western mountain to the earth-surrounding ocean. This mystical river was personified in Anahita, who appeared "in the shape of a maid, fair of body, most strong, tall-formed, high-girded, pure, nobly born of a glorious race" [8]. Through the Persian conquest of the Middle East Anahita later became assimilated with the local goddesses such as Ishtar and Aphrodite. After the fall of the Persian Empire, the cult of Mithra and Anahita was adopted by the Greeks, who settled in the Levant, to be followed later by the Romans.

Thus the ancient goddess of fertility symbolized by the Iranians as a holy river springing out from the earth and the ancient sun god or the god of light symbolized by the sun's daily resurrection from the dominion of darkness survived and defeated all attempts at abstraction and monotheism. They continued to rule the minds and hearts of man for many generations to come. The destruction of the Persian Empire came by the hand of the Greeks, but its deterioration was caused by the Persian rulers, who indulged in a race for power and luxuries.

In the beginning, when the Persians emerged from their desert land which they had turned into a fertile country by their ingenuity, diligence, and righteousness, they swept over the Fertile Crescent and the Anatolian peninsula. They conquered Lydia and subdued its King Croesus, who was a symbol of richness and pride. The fall of Lydia brought the Ionian cities (Polis) in western Anatolia or Asia Minor under the rule of Cyrus the Great. In 499 B.C. these cities revolted against Darius. Greek ships from Athens and Eretria helped the rebels. Thus, after subduing the rebelling Ionian cities in Asia Minor, Darius sent his army to the Aegean peninsula, where it was defeated by the Athenians at Marathon. Darius did not try again to pit his army against the Greeks, although he did not hesitate to invade the vast plains of southern Russia or Scythia, north of Greece. He was forced back, not by the Russian winter, but by the lack of water for his great army. He withdrew, leaving behind the sick and the stragglers, without being able to inflict a decisive blow on the Scythians, who adopted the warfare of sudden attacks by their horsemen, evading full-scale battle.

Ten years elapsed before the Persians, under Xerxes, tried again to conquer Greece. During these ten years the Greeks struck a rich vein of silver in their mines, with the profits of which they built a navy. When the Persians attacked again they were more successful in the beginning. They advanced and conquered all northern Greece, sacking and burning Athens. This war was decided by the Greek navy, which dealt a defeating blow on the Persian navy composed of Phoenician and Ionian ships. In the next summer, the Persian army was defeated on land by a Greek coalition, with the Spartan army as the elite troops. What was won from the Persians by the unity of the Greeks was lost during the many wars which flared up between the Greek cities. These internal wars weakened all of them and enabled Philip, the King of Macedonia, a semi-civilized territory in the northern part of Greece, to gain power over and control the main Greek cities. The son of Philip crossed the sea back into Asia, defeated the King of Persia, and became the master of the Persian Empire.

12 Water, Monarchs, and Martyrs

> Now in the fifteenth year of the Reign of Tiberius Caesar, Pontius Pilate being governor of Judaea, and Herod being tetrarch of Galilee, and his brother Philip tetrarch of Ituraea and of the region of Trachonitis, and Lysanias the tetrarch of Abilene, Annas and Caiaphas being the high priests, the word of God came unto John the son of Zacharias in the wilderness. And he came into all the country about Jordan, preaching the baptism of repentance for the remission of sins. (St. Luke, 3:1-3)

The victory of Alexander over Darius decided the fate of the Fertile Crescent for 1000 years; it also decided the faith of what is called the western world for 2000 years.

The first counter-impact from the eastern culture on the west started the moment Alexander the Great visited the shrine of Amon in the desert of Egypt, where he was greeted by the high priest as the son of Amon-Zeus. This was immediately followed by the glory bestowed upon him by his decisive triumph over Darius. Then came all the luxuries and pomposity of the Persian Court which became his inheritance after he conquered Persia. The paganish heritage of the Hellenes of an immortal hero, a bastard son of gods, like Hercules, was grafted onto the belief of the eastern world that kingship is divinity and the king is the god's representative upon earth. This composite world view following the comet-like appearance of Alexander, the Macedonian, in the skies of the civilized world, haunted the western world for many centuries to come. The heritage of Greek democracy, Roman republic, Christian humility, French egality, and German enlightenment were to be overpowered and shattered again and again by the dreams of many men wanting to be another Alexander and to decide the fate and faith of the world.

According to the Jewish historian of Roman times, Josephus Flavius [1], Alexander, during his siege of Tyrra, one of the main naval bases of the Persian Empire, sent a mission to Jerusalem demanding its surrender and the joining of the Jews to his army to fight the Persians. The high priest turned down this demand, replying that he was under oath of allegiance to the Persian king. After Alexander conquered Tyrra and Gaza, he turned to Jerusalem with the purpose of punishing this city for its disobedience. The high priest dressed himself in his robes of office and came towards the marching Greek army. When Alexander saw the high priest he dismounted from his war chariot and prostrated himself before him. To his astonished lieutenants he explained that he had seen the image of this man when he embarked on the campaign against Darius and the man had promised him victory. According to the same story, Alexander then went with the high priest to the temple of Jerusalem, ordered a sacrifice to God , rendered religious autonomy to the Jews. He even exempted them from taxes each seventh year, the year of "Shemita" during which the Jews left all their fields uncultivated. We do not know, of course, what is the factual truth behind this story. Did Alexander really prostrate himself before the high priest? If he did, was it because he had seen the man in his dream or because he thought that such a gesture and story would secure his rear, as well as the goodwill of the Jewish people in Babylon and Persia? We have to remember that he had still not finished the war, and was on his way with his army eastward to fight the remaining Persian army. What we do know is that, although he was not proclaimed a son of the God of the Jews, yet Jerusalem and the temple inside it were saved, and religious as well as political autonomy, granted in the past by the Persian kings, was maintained by the Jewish people living in Judea. From Jerusalem, Alexander went to Egypt where he visited another temple, was proclaimed a son of god and did all the deeds we know from the history books. From that historical moment onward the dominance of the east by people of Hellenic origin began. This also brought about the creation of the Hellenistic culture, a hybrid of Oriental and Greek paganism, sprinkled with Greek philosophy and adorned by Greek architecture and art. At the same time, the great conflict between Hellenism and Judaism, namely between paganism and monotheism, began, a conflict whose resonance continues to reverberate in the world even today.

Although Alexander adopted the oriental concept of kingship, yet the Greek way of life had a dominating impact on the Orient, especially in the countries in which the dynasties of Hellenic origin survived for longer periods. Alexander's heirs, like all followers of people of special distinction, knew very well how to obey and deliver the commands given to them, but did not possess the genius to overcome personal conflicts and strive for a universal goal. Thus, after Alexander's death, his kingdom was split between his followers amid a series of cruel wars of one part of the army against the other.

Egypt, Palestine and Syria fell under the rule of Ptolomeus, while most of what remained of the Fertile Crescent, including Anatolia, fell to Seleucus. Macedonia remained in the hands of Antigonus, grandson of Alexander's general who was killed when the wars between the generals started.

Although Ptolomeus and Seleucus ruled their pieces of Alexander's empire as oriental absolute monarches, much of the Greek way of life and freedom of thought was maintained and attained the unique achievement of the Hellenistic period. Ptolomeus became a new Pharaoh, but he gave to three large cities the rights of a

Greek polis, the greatest and most famous of which was Alexandria, established by Alexander when he conquered Egypt. On the other hand, all Egypt was ruled according to the Pharaonic system in which the peasant and the simple citizen were no more than worshipping and tax-paying slaves to their king-gods.

It was thus a thin veneer of a ruling class, centered mainly in Alexandria around the Ptolomean court, who lived at the expense of the perpetual, obedient, Egyptian peasant oppressed by the tax collectors, a system which had characterized Egypt since its early history. Yet, due to Greek maritime traditions and way of life, as well as because of its place between Asia and Africa, Alexandria became the center of commerce of the world. The Ptolomeans built a huge fleet, which mastered the Mediterranean, the Red Sea, and the Indian Ocean. It also ensured a safe passage from Greece to Egypt, from where the Ptolomeans drew their mercenary army. The safe traffic to the mainland also enabled close ties with the scholars of Greece and the Greek colonies. The establishment of the museum and library, and the financial support which the Ptolomeans were wise enough to guarantee to the scholars who taught and did research in these institutes, brought about the flourishing of science, technology, and literature. Although the center of science in Alexandria can claim original contributions to science, such as that in mathematics connected with the work of Euclid, astronomy with the name of Hipparchus, and geography with that of Eratosthenes, yet in the study of botany and zoology, Aristotle's Lyceum in Athens, while he was alive, was the leading group. This school amassed a tremendous quantity of observations on all aspects of nature. Their explanation of natural processes was bounded by the doctrine of the "four elements" from which all material bodies are built: fire, air, earth, and water. The school of Aristotle influenced the thought of the Middle Ages and, while the Bible was believed to contain all the truths regarding God and His wishes, Aristotle's works, or what has survived of them, was believed to contain all the truth regarding nature.

To these works we owe the blocking effect to the direct approach to nature which characterized European science during the Middle Ages . As a result, it brought about also the misunderstanding of the hydrological cycle, especially the connection between surface and subsurface or groundwater. According to Aristotle's explanation, springs did not arise just from precipitation infiltrating into the subsurface, but also from vapor which condensed inside the earth's crust, as well as vapors from some other source (juvenile water, most probably). These views were investigated and explained by generations to come, nobody daring to question the master's explanations until Leonardo da Vinci (1452-1519) in the Renaissance and, a little later, Bernard Palissy (1514-1589), who challenged Aristotle's doctrine. The coup de grace to the master's hydrological speculations was dealt by Pierre Perrault, who, in 1674, proved that the flow of the Seine is equivalent to the quantity of the precipitation on its drainage basin. (For the detailed story of the evolution of the scientific explanation on the origin of springs, the reader is referred to [2]). It is interesting to note that Aristotle's explanations were based on observations and as a matter of fact, that the theory explained in the most rational way the facts known to science at that time. For instance, the fact that in the karstic caves of Greece and

Ionia the roofs of the caves dropped water all year long, even during the hot, dry months of summer. Another observation relates to the submarine springs which during high tide may form conduits for sea water to flow inland. One also has to take into consideration that an explanation was needed for the thermal springs found everywhere in this tectonic active belt. The harm to man's progress in understanding nature was due, thus, to the following generations, who turned a hypothesis into a dogma, the reason for this being the debate and later clash between the followers of a new religion, namely Christianity. The sect which won the debate decided that the only way to prevent any future objection to its hegemony was by fossilizing all ideas and by censoring any innovation to the once accepted dogma. The theory explaining the origin of springs had to wait even longer than the heliocentric theory in order to become an acceptable explanation.

The smaller part of Alexander's empire consisted of Macedonia and Greece, and was left in the hands of Antigonus II. In 280 B.C. it was attacked by hordes of Celts or Gauls, who penetrated the Balkan peninsula, reaching Delphi in the south. Antigonus defeated the Celts and established the Macedonian empire.

This invasion by the Celts took place at the time when the curve of the ^{18}O (Fig. 1.2) shows a pronounced depletion which the author interprets as a sign of the start of a cooler period (see also Chapter 1). This, in his opinion, affected the plains of central and northern Europe and was the cause of the southward migration of the Celts.

At the same time the Nabatean nation started to emerge in the desert plains of northwestern Arabia, Sinai, and Negev. These people, starting from dispersed nomadic tribes of Semitic origin, settled in the Negev and later also around a center in southern Jordan, called Petra. Towards the end of the 1st century B.C., they established a strong kingdom, the richness of which was based on agriculture and commerce [3]. It is the opinion of the author that these two events, though remote in space, are interconnected. The invasion by the Celts from the north was a desperate effort to escape the hazards of the cold and try to settle in the warmer Mediterranean countries, while the settling of the desert nomads and the development of a sophisticated system of agriculture based on water harvesting , which will be discussed later, was due to the desert becoming more hospitable. The author suggests that the two events were the results of the beginning of an advance of the glaciers which can be even termed "a mini glacial" period.

The rule over the central part of Alexander's kingdom fell into the hands of Seleucus, who adopted Alexander's plan of transplanting the Greece way of life into Asia. Seleucus and his son, Antiochus I, founded Greek polis-type cities, inhabited by Greek immigrants, throughout their kingdom. The government of these cities was based on that of the Greek polis of a semi-autonomous, free community in all that referred to internal affairs.

The war between the heirs of Ptolomeus and Seleucus was centered on the rule over Judea and Edomea, the land which to the Seleucids meant a vital part of their heritage, and to the Ptolomites meant control over the western Mediterranean. This land passed from hand to hand until 200 B.C., when Antiochus III triumphed over the armies of Ptolomeus IV in the battle at Banias, the spring at the headwaters of the Jordan. This brought all the western Mediterranean, including Judea, under the rule of the Seleucids.

The same Antiochus III suffered a severe defeat by the Romans in 190 B.C. He had to pay heavy indemnities to the Roman republic, which caused a severe economic crisis in his kingdom. He himself was killed while trying to confiscate the treasures of a rich temple in Elam. His son, Seleucus IV, tried to do the same in the temple in Jerusalem but was repelled.

Until this time, the Jewish people in Judea and the temple in Jerusalem had maintained their autonomous religious and administrative rights, under the Ptolomites and later under Antiochus III. This situation changed when Antiochus IV, or Epiphanes, became king. In repulsing an invasion by the Ptolomites, he invaded Egypt and nearly conquered Alexandria. The intervention of Rome forced him to retreat. On his way north he pillaged the treasures of the temple of Jerusalem. In order to secure the loyalty of the inhabitants of Judea, he started an intensive political campaign to promote Hellenism among the Jewish inhabitants of this province. This involved giving special privileges to those who adopted the Hellenistic way of life, and at the same time prohibiting the Jewish cult and circumcision. This brought about the rebellion of the Jews which, after a few successes and defeats, led to the establishment of an independent Jewish state under the rule of a family of priests who were the initiators of the rebellion. This family established a dynasty of kings-chief priests by the name of the house of the Maccabees or Hasmoneans.

During the low ebb of the war of independence, the Hasmoneans escaped to the desert in the eastern part of Judea where they found refuge from the Greek army. This region, which is in the shadow of the rainstorms coming from the Mediterranean, receives less than 200 mm annual precipitation. Due to its proximity to the Dead Sea, which is a young geological feature, the region is dissected by deep canyons. The rugged topography and dryness make movement in this region very difficult enabling hiding and reorganization. Later, when their rule over Judea had been established, the Hasmoneans built many forts all over the desert, and provided these forts with elaborate canal systems by which water was collected from the surrounding hillsides, to flow into huge cisterns, the capacity of which enabled these forts to withstand siege for months and even years. When Trans-Jordan was conquered, other forts were built on the east of the Dead Sea, too. These forts were used later by the heirs of the Hasmoneans, who fought each other for the right to the throne.

These wars between the brothers made it easy for Pompeius, the Roman, to extend the rule of Rome over this part of the world. He was asked by one of the heirs to help him fight against his brother.

These wars, and the wars between the Roman leaders, enabled a general by the name of Herod to seize the throne and become king. Herod was from Edom, a nation in southern Judea. The Edomeans had been forced to become Jewish after they had been conquered by one of the Hasmoneans kings. Herod married a daughter of this family, whom he later killed when he became suspicious that she had betrayed him.

Herod gained power by his shrewd manipulations in the Roman court and by knowing how to gain the good will of the Jewish population who were weary of the wars between the Hasmonean heirs. His main success was when he changed sides in time and backed Octavianus against Antonius, the patron of Cleopatra, who was one of Herod's chief enemies.

In order to become popular with the Jewish population, Herod distributed food from his granaries when a few dry years struck the region. After he became king, he started to annihilate the Hasmonean family. He invited his brother-in-law, Aristobolus, who was also the high priest, to the palace in Jericho which had been built by the Hasmonean kings and was furnished with baths and swimming pools, and there, by his order, Aristobolus was drowned .

When Antonius was still in power, Cleopatra received from him Jericho and the palaces as a gift. After Cleopatra and Antonius committed suicide following their defeat by Octavianus, Herod crossed to the victor's side and regained Jericho, its gardens, and its palaces.

Being constantly in fear of a rebellion by the Jews, Herod enlarged existing, and built new, desert fortresses. One of these was Herodion, east of Jerusalem, which he had chosen also as his burial place. Another was Massada, on one of the tilted limestone blocks along the main fault line bordering the Dead Sea on the west. A third was Makhwar, on the eastern side of the rift valley. These two fortresses were the last to stand against the Romans at the end of the big Jewish rebellion. Herod built not only fortresses, but also cities, temples, ports, water supply projects, amphitheaters, etc. In fact, he was the greatest builder in history that this part of the world has ever known. In honor of his patron, Octavianus Augustus, he built two new cities which he named after his benefactor. Sebastey, in the mountainous part on the place where the ancient capital, Shomron, of the kingdom of Israel stood, and Caesarea, on the coast. The harbor, walls and other utilities were the most magnificent the Roman empire had ever seen. Of special interest to us is the aqueduct (Fig. 12.1) which brought water from the springs flowing from the limestone aquifer of Mount Carmel, about 15 km northeast of Caesarea. The city had a sewage system, hippodrome, and an amphitheater in which gladiators and wild beasts killed each other to celebrate the dedication of the new city to its overlord. The other large building was the great temple of Jerusalem., which he built in order to please the Jewish population in Judea and also that scattered all over the Roman world. This was also an outlet for his unquenchable desire to build and leave his name for generations to come.

Jerusalem was, at that time, the religious center for the entire Jewish world, which spread throughout the Roman Empire and beyond. The temple was the center of Jerusalem. During three Jewish holidays, Passover, the Feast of Tabernacles (Succot), and Pentecost (Shavuot), Jews from everywhere came on pilgrimage to Jerusalem. In order to supply water for these pilgrims, Herod built a gigantic water supply system which was fed by all the springs in the vicinity of Jerusalem, as well as those on the road to Hebron to the south. The project starts about 25 km south of Jerusalem, yet the canals and conduits are 40 km long, and because of the difference in altitude between the source and outlet, the gradient is 1:1000. This project was undoubtedly carried out with the help of Greek or Roman engineers. The special ingenuity that they showed in this project was the utilization of all the small springs and the way they overcame the problem posed by the many streambeds (wadis) which intersected the course of the water conduit. Along it there are large reservoirs to guarantee a continuous supply, even when a part of the water conduit became damaged. The system draws from two sources, each collecting water from a few springs. The more western branch starts with a tunnel hewn in the limestone of Cretaceous age. The tunnel has vertical shafts similar to those of the ghanats (described in Chap. 11). A similar structure in solid rock is found in Italy and was built by the Etruscans. Whether the system in southern Judea was based on Etruscan engineering methods transferred to those of the Romans or Persian, is difficult to say.

South of Bethlehem the two branches converge into a system of three large reservoirs, the popular name of which is King Solomon's Pools (Figs. 12.2,12.3). Two of these were built mainly during Herod's time, while the third was built during the period of Turkish rule. From these pools, a branch led to Herodion, Herod's magnificent fortress, desert palace and mausoleum. Two other conduits lead to Jerusalem. One of them crosses Bethlehem by a tunnel. The other conduit, on a higher level, crosses a valley near Bethlehem by a syphon built of stone sections which, when put together, form a pipe (Fig. 12.4). This part of the system was rebuilt partly by the Roman Governor of Judea, Pontius Pilatus, and later by the Roman army that vanquished the Jewish rebellion.

Josephus Flavius [1, 5] reports that Pontius Pilatus confiscated the treasures of the temple in order to dig a water canal to Jerusalem This irritated the Jews and caused riots. This was additional fuel to the cinders of rebellion which were already burning in the hearts of many Jews. On some of the sections of the aqueduct in Bethlehem the name of the 10th Roman Legion, that were stationed in Jerusalem after the rebellion was crushed is found. Inside the City of Jerusalem, subterranean reservoirs as well as open pools were built and reconstructed. Some of them maintain water even today. Some, like the pool of Bet-Hesda, have been filled by debris over the centuries. These reservoirs provided a safe supply of drinking water to the inhabitants of this big city and to the multitudes who came as pilgrims to the temple at its center.

The reservoirs also furnished enough water for the many special ritual baths which were in use in Jerusalem at that time. The ritual of washing in these special baths or "Mikvas" was a very important part of the ritual of pilgrimage, as a Jew was not supposed to enter the yard of the temple without going through the ritual of purification in a "Mikva". As we will see later, the ritual of purification in water was an important part of the religious ceremonies of some of the Jewish sects that flourished during the time of Herod and his heirs' in Judea.

The temple, which was rebuilt by Herod, was one of the most magnificent buildings of the ancient world. The temple itself was erected on an outcrop of limestone (Turonian-Middle Cretaceous age) which, according to Jewish tradition, was the site where the Patriarch Abraham intended to sacrifice his son, and also the cornerstone of the world. The yards of the temple were built on an artificial platform supported by massive walls, the western wall of which survived the destruction of the temple by the Romans. For the Jews it remains a symbol of their past glory and for about 2000 years they have come to this wall to wail over the destruction of their temple.

Fig. 12.1. The Roman aqueduct at Caesarea (Photo courtesy Irit Zahroni, Bamahne Journal, IDF)

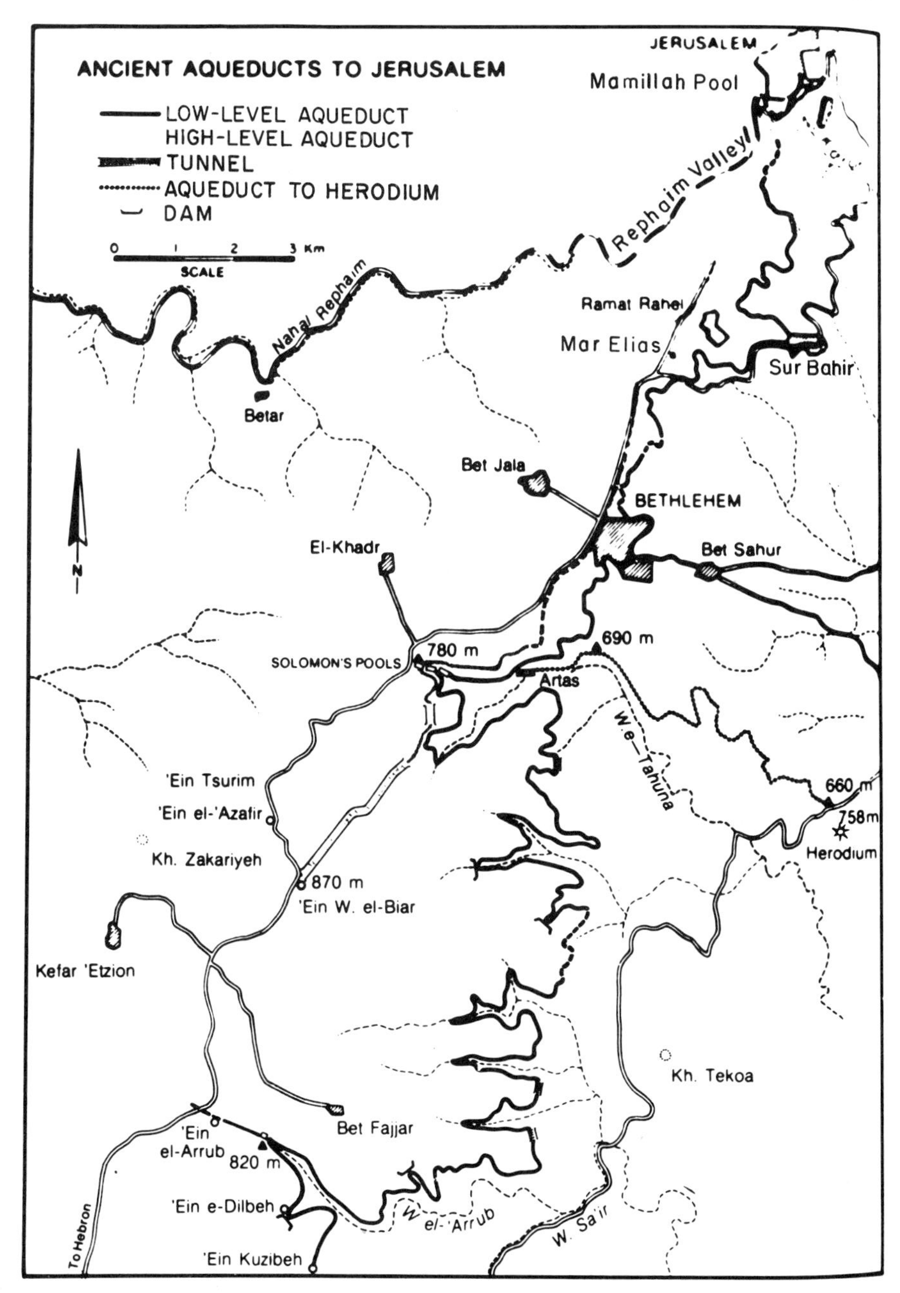

Fig. 12.2. Map of aqueducts' system, Jerusalem (Courtesy M. Harel, University of Tel-Aviv, see General Bibliography)

Fig. 12.3. "Solomons Pools" near the Hebron-Jerusalem road (Courtesy Irit Zaharoni, Bamahne Journal, IDF)

In order to reduce the pressure of the platform on the supporting walls, Herod did not fill in the inner part of the platform, but left it empty, supporting the floor of the yards of the temple by an elaborate system of arches. Part of this subterranean space was utilized for water cisterns, which are still in use today by the pious Moslems who pray in the mosques that were built on the platform after Jerusalem was reconquered by the Moslems from the Crusaders. The latter, who did not attribute any holiness to the temple site, used the yards as an army camp and the arched subterranean space as stables for their horses. As in the case of the water supply project built by Herod, they decided that only King Solomon, the wisest of all men, could have built such a structure and so named it "King Solomon's Stables".

Josephus Flavius reports [1] that for the building of the temple Herod trained a thousand priests as artisans. We do not know exactly what was the reaction of these priests when required to leave their traditional occupation of slaughtering animals for sacrifice and praying for the welfare of the donors, yet what has survived of these buildings shows that the clergy, when put to constructive work, can be rather useful.

Besides the ordinary ritual of bathing in a Mikva, which every pilgrim had to perform, there was a special ritual for libation of water which was performed during the Feast of Tabernacles. This feast of seven days was celebrated in the autumn and special prayers were said during the festivities asking for sufficient rains to water the fields of the land. While performing these prayers, they also carried in their hand a branch of the palm tree, the willow, the myrtle, and the fruit of citron, apparently symbols of plenty of water and fruits. While during the other festivals throughout the year the libations on the altars of the temple consisted of wine and olive oil, during the Feast of Tabernacles they consisted also of water, which was brought from the spring of Siloam which, as previously explained, was connected by the "Niqba" to a pulsating karstic spring "Gihon". This was used also as the site for the enthronement of the kings of Judea during the time of the First Temple. The water was taken from the spring on the first night of the holiday in a vessel of gold carried by a priest, who then sprinkled the water on the altar amid festivities which included the lighting of oil lamps and dancing with torches.

Fig. 12.4. Pipe made of stone crossing Bethlehem (Courtesy Irit Zaharoni, Bamahne Journal, IDF)

The libation of water accompanying the prayers for rain was undoubtedly a Canaanite tradition adopted by the Jews. It is not mentioned in the Pentateuch, where all sacrifices and libations are listed. For this reason, the party or sect of the Sadducees, who voted for a conservative adoption of the commandments written in the Bible, were against this libation. On the other hand, most of the people who backed the sect of the Pharisees, who maintained that the commandments of the Bible could be interpreted and thus changed by the rabbinate, adopted this festival of the drawing of the water. One of the Hasmonean kings, who were also high priests, probably Alexander Janai, favored the Sadducees, and thus, when asked to perform the libation of the water, poured the water on his legs instead of the altar. He was stoned with the citrons, which the Jews happened to be holding in their hands, part of the ritual of the holiday.

We do not know whether this water libation or other reasons brought about a period of years of plenty during the Hasmonean kingdom (166 B.C. to 36 B.C.), Herod's period (37 B.C. to 4 B.C.), and later during the Roman Empire, until about 300 A.D. The author's opinion is that it had to do with a more humid period which was due to the mini-glacial period . These years of plenty can be witnessed in the sediments of the Sea of Galilee where, during the same period in which there occurs a depletion in the ratio of ^{18}O and ^{13}C isotopes, there also occurs an increase in the ratio of olive pollen and reduction in that of oak. As mentioned earlier, fluctuations are explained by the cutting of the natural vegetation and planting of olives as a result of the abundance of rains which enabled the expansion of the olive groves. We do not know what might have caused this climatic change. It might have been an extra-terrestrial reason, such as an increase in the number of sun-spots, which could have caused the reduction in solar radiation. This reduction might have become more severe due to volcanic dust thrown into the atmosphere by volcanic explosions. The ice cores from Greenland [5] show us that there were two rather severe eruptions in about 250 and 200 B.C. Another, more severe one, occurred in about 50 B.C. The Roman Virgil tells that "when Ceasar died (44 B.C.), the sun felt pity for Rome, as it covered its beaming face by darkness and the impious generation feared an eternal night" [5]. A century later, the elder Pliny also mentioned the weakness of the sun when Caesar was killed . In Judea, the level of the Dead Sea rose to about 70 meters above its present level [6], (Fig. 1.3). The level of the Nile was lower during this period [7], (Fig. 1.3). In England, many Roman buildings built prior to 300 A.D. were covered by the transgressions of the sea which occurred just after this date [8], an indication that the sea retreated when these buildings were built.

All these data show that, from about 300 B.C. to about 600 A.D. [with a short warming-up phase around 300 A.D. (Fig. 1.3)] there occurred a mini-glacial period which caused an advance of the ice in the polar regions and high mountains, a regression of the sea, a colder climate in northern Europe, and a more humid climate in the deserts of the Fertile Crescent. The regions benefiting from the monsoons, such as the Ethiopian highland, the Sahel, southern Arabia and maybe even India, most probably suffered from drought, as indicated by the rather low levels of the Nile. The more benign climate in the desert of Judea and Edom, in the opinion of the author, promoted the flourishing of the Nabatean kingdom, which was established in the first century B.C. in the Negev by nomad tribes, apparently of Arabic origin. These tribes controlled the trade routes, especially that of the frankincense and myrrh, which was brought from southern Arabia to the Mediterranean ports. Their capital city, for some time, was in Petra, situated in a canyon transversing the mountains of Edom. In the desert of the Negev, they built six beautiful cities (Figs. 1.1, 12.5), which flourished until the 7th century A.D. Two main factors attributed to the prosperity of the Nabatean kingdom. The first factor was the trade routes they controlled and the second, the sophisticated system of irrigated agriculture they developed. The desertion was attributed by most investigators, as referred to in Chapter 1, to the Arab invasion. In the following paragraphs the hydrological aspects of the Nabatean settlements in the desert will be discussed (main sources [4, 9]).

The system was based in principle on the efficient use of flood water . The water was harvested mainly from small watersheds, the area of which ranged from 10 to 100 hectares spread over the slopes of the hills surrounding a terraced valley in which the agricultural farm was situated (Fig 12.6). The water was collected from the slopes by channels with a slight gradient, which led the water to the terraces in the river-bed. The latter were built of stone, with spillways to let any surplus of water pass from the upper terraces to the lower ones. A few collecting channels, one parallel to the other, divided the hillslopes into rather small plots, about 1-1.5 hectares each, the water from each being led to the terraces in the valley. This subdivision increased the efficiency of the water-harvesting, as the smaller the area the higher the ratio of runoff to rain. The ratio between collecting hillslope area and valley farm area was on the average 20:1, namely that each hectare of farmland necessitated around 20 hectares for its supply of irrigation water. The team which investigated this type of agriculture [4, 9] did not consider any climatic changes, and assumed that the rate of precipitation was of the same yearly average as at the present, namely 100 mm. Thus for an average of 15% runoff, each hectare produces on the average 150 m^3 of water, 20 of which will produce about 3000 cubic meters. In the opinion of the present author, this quantity is sufficient to sustain the plantations, but not to guarantee good yields. In his opinion, the quantity of rain was 50% more than the present, which means a yield of about 4500 m^3 for each hectare of farmland, which is the correct amount to ensure good yields. This, together with the fact that there were apparently less years of drought, may explain the fact that the ancient inhabitants of this arid region did not neglect even the smallest river-bed to be terraced and farmed. They did not go beyond the present iso-rainfall line of 50 mm.

In conclusion, the opinion of the author is that the many terraces north of this iso-rainfall line (which at that time was the 75 mm line), the large number of wine and oil presses and wheat mills, suggest that the richness of the Nabatean cities did not stem solely from trade but also from agriculture. The conformity between the isotope-pollen curves (Fig. 1.3) and the flourishing of the Nabatean cities is not accidental, namely it means that the reason for this prosperity was due to a more humid climate. The end came slowly when the climate became warmer and dryer. When this happened, the sophisticated water-harvesting methods did not help, the terraced fields had to be abandoned. It should be added that these cities were not destroyed, but were deserted. The contemporary visitor can still walk in their streets and visit their beautiful churches. To his question, what caused the desertion, man or nature?, the author answers: the first cause was nature, which triggered a process which man later accomplished. The process on the whole is that of desertification.

One of the puzzles of the Nabatean agriculture were the small stone mounds which covered the slopes of the hills around the farms. The local bedouins called them "Tuleilat el enab", which means the mounds of the grapes. Some claimed that they were built for trellising the vines, while the stones caused the condensation of dew, as supplemental irrigation. Prof. Evenari and his colleagues [4, 9] maintain that this bedouin name has nothing to do with the real purpose of these mounds. In their opinion, they were meant to increase the harvest of water from the slopes, by making the surface of the hill slopes smoother. They showed by experiment that such a treatment of stone collection to a slope renders about 25% more runoff water than from a slope on which the stones were left strewn.

Fig. 12.5. Air photo of the citadel of the city of Avdat. In the wadi in the background the reconstructed ancient farm can be seen (Photo courtesy Irit Zaharoni, Bamahne Journal, IDF)

The present author suggests, in connection with his recent findings regarding a climatic change and infiltration into the rock fractures, reinvestigating the possibility that these mounds were indeed used for vines, while the additional runoff was an added benefit. Another method of irrigation was diverting floodwater from a large wadi, by the aid of a diversion dam. The water covered ancient terraces caused by the river in the wadi, which, in addition, were artifically terraced by stone walls. Such projects needed large stone structures to withstand the ferocity of the floods of the desert. They also needed continual maintenance, such as cleaning sediments which became deposited in them. Neglect, as in the case of the big dam of Mamshit (Kurnub), caused them to become filled to the top by sediments, as well as breached as in the case of Ein Qudeirat [10]. In order to supply drinking water, the Nabateans collected water that drained from the roofs and pavements of their cities into large cisterns. Outside the cities they excavated large chambers in the chalky rocks which served as cisterns. The water was diverted from the wadi bed by a diversion channel through a sediment trap. These cisterns were used, for watering herds, and might even have helped in the supply of drinking water to the cities. In these cave-like structure evaporation during the summer months was minimal.

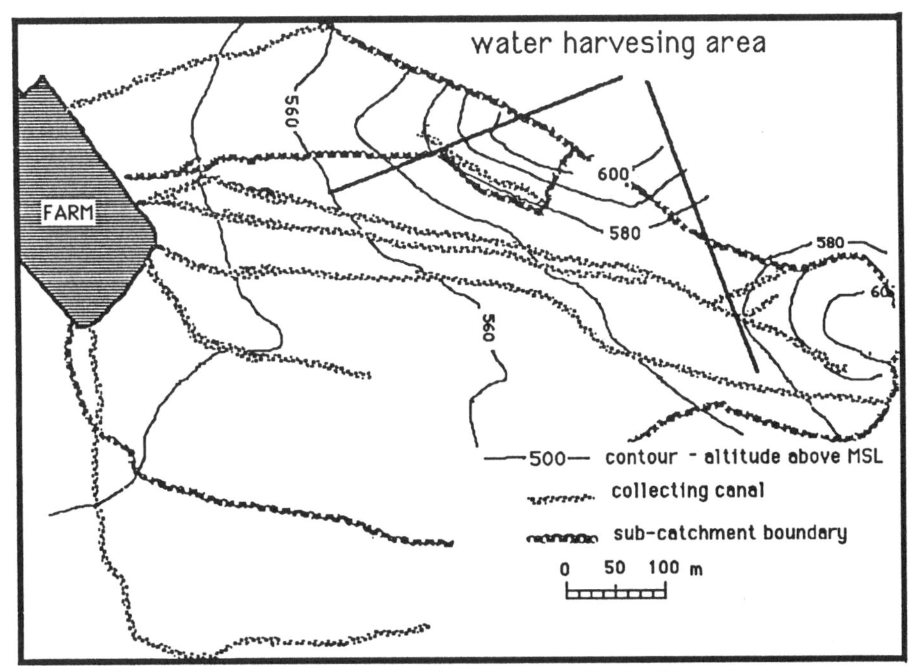

Fig. 12.6. Plan of ancient water harvesting system at Avdat [9]

In the cities of Nizzana, Avdat, and Rehovot, deep wells were excavated into the chalks of the Eocene age for locating groundwater. The introduction of this system occurred after the Nabatean kingdom was annexed to the Roman Empire in 106 A.D. The investigations carried out by the present author revealed that the ancients were familiar with the system of groundwater flow in chalks. Thus, while still knowing that the chalks are in general soft and impermeable and may be utilized for the construction of cisterns for storage of water, they discovered that in the vicinity of larger fractures the water infiltrated to great depths to form the groundwater table [11]. They excavated wells to a depth of about 70 meters in order to tap these fractures and promised a safe yield of water for their cities. In Avdat they used brackish water (900 mg/l Cl.) found in such a well (60 m deep) to supply a Roman bath furnished with all the necessary facilities.

The more humid climate may also explain the flourishing of the "Decapolis", ten cities in Trans-Jordan, yet this subject needs more investigation.

The Judean Desert was still rather desolate region due to its climate but mainly to its rugged topography, yet, when precautions had been taken to secure a water supply, the desert could be an ideal place for people seeking refuge or solitude. A special type of a community was established from ca.150 B.C.to 70 A.D. by a sect that took advantage of the desert as a place of solitude and study. This was the Sect of the Dead Sea Scrolls (which most scholars believe to have been the Essenes), who built a center for study near the Dead Sea, at a place called Qumran. This place was most probably called by the Essenes "Mezad Hassidim" (The Stronghold of the Devotees). In this place, ancient scrolls closed in pottery jars were found. When opened and read, they struck the historians of the Western World with astonishment, as these were the writings of a Jewish sect from the time of the Second Temple. Besides known parts of the Bible, such as the prophecies of Isaiah, there were also scrolls which told about special rules of purity which the members of the sect imposed upon themselves. The many ritual baths or "Mikvas" found at Qumran, supported by an elaborate system of water collection and storage (Fig. 12.7) are evidence of the importance which these people attached to the ritual of bathing.

It is not that the other Jewish sects did not attach importance to bathing as a purifying ritual. On the contrary, as was stated earlier in this chapter, the ritual bath was one of the foundations of the laws of purity among the Jews in the temple and at home. One finds the obligatory commands for this ritual in the books of the Pentateuch, especially in the book of Leviticus. That these commands were obeyed is obvious from the many "Mikvas" found in archeological excavations in Judea, especially in Jerusalem, from the time of the Second Temple (515 B.C. to 70 A.D.). Yet the sect of the people who wrote the Dead Sea Scrolls put special emphasis on this ritual, as they saw themselves as a distinct sect chosen from among all the other Jewish sects, whom they regarded as relatively impure. The disassociation of the Essenes, to become a special sect of ascetics, was most probably during the later period of the Hasmonean kingdom, when social injustice and the adoption of Hellenistic ways of life which accompanied the moral deterioration of this dynasty became apparent. At that time, the sect of the Sadducees was favored by the ruling family and nobility, while the multitude supported the Pharisees. The Essenes believed that these two sects had abandoned the pure and true Jewish way of life. They separated themselves from the others by obliging themselves to carry out the very strict laws of a community which strived to bring upon the world the rule of the light which, for them, was the dominion of God over the Jewish people as the Chosen People. The scroll of "The War Between the Children of Light and the Children of Darkness" describes the war and triumph of the Children of Light at the Day of Judgement. The light and darkness may be an exponent of a dualistic motive in the religion of the sect.

A man named Johanan, known today as John the Baptist, who was, most probably, one of the followers of this sect (or a subsect, the hemerobaptists), also prophesied the coming of the Day of Judgement and called the people to purify themselves in the water of the Jordan. He was admired by the local people. He also preached for the distribution of the riches of the people among the poor and the adoption of a life of nonviolence. Johanan's preaching, the admiration the people had for him, and the social unrest which followed Herod's death caused him to be hated by the ruling class. When Herod Antipas, the son of Herod, fell in love with Herodias, his step-brother's wife, arranged her divorce, and married her, he was reproached by John the Baptist. Herod Antipas had John the Baptist arrested, brought to Makhwar, the desert fortresses in Trans-Jordan, and there executed.

One of Johanan's followers, named Joshua, known today as Jesus, preached in Galilee, the province of his birth, a similar message to that of John the Baptist. He told the poor people about the Day of Judgement which was soon to come and the need to become pure and humble in order to avoid punishment. He promised them salvation and a kingdom of heaven. Their sufferings, he pointed out, were a virtue. He was thus popular with the poor people and hated by the rich. He did not agree with the way the Pharisees and Sadducees interpreted the Jewish laws, and demanded that more emphasis be put on the laws of justice than on those of sacrifice and ritual. He came to Jerusalem and infuriated the priests of the temple by overturning the tables of the money-changers in the court of the temple. He was arrested and brought before Pontius Pilatus for trial. When the Roman found that he was a Galilean he turned him over to Herod Antipas, who was the ruler of this region. He interrogated him, but receiving no replies, returned him to Pontius Pilatus, who sentenced him to one of the most cruel punishments in the Roman Empire, specified for rebels, that of crucifixion.

Johanan's and Joshua's preaching continued to spread after their martyrdom. It started in Galilee and Judea, and spread all over the pagan world. In the beginning it was just another Jewish sect which preached for reformation of the established religion and the renovation of the social and humanistic values of Judaism. Its followers were mainly from the masses, but some of the Pharisees backed it.

Most of the Pharisees, the Sadducees, who were with the ruling establishment, the heirs of Herod and the Romans, not to speak of the rich people, persecuted the followers. The persecutions caused these people to look for refuge outside Judea and Galilee. They preached their ideology to the Samaritans and the non-Jewish people who lived in the Hellenistic cities. The monotheistic, yet humanistic, contents found many supporters, especially among the enlightened people who were growing away from the views of the pagan world. This was in step with a general trend in the Hellenistic-Roman world to adopt the Jewish faith. Many people in the Roman Empire found in the Jewish monotheism answers to their quests and converted to Judaism, while many others adopted only a part of the Jewish laws. It was Saul, or Paulus, of Tarsus in Asia Minor, who saw how Christianity could take over this tide of monotheism. He recommended abolishing the main laws which, until then, distinguished the Jews from the non-Jews, such as the laws of circumcision, taboos of food, and others. One of the Jewish laws that was dropped and adopted only in a symbolic way was baptism. The non-Jewish Christian, from that day on, could do without the many ritual cleansings a Jew or Jewess had to take after he relieved himself or before taking food. Symbolic baptism in infancy seemed to suffice, especially in the cold climate of Europe, to where Christianity spread. However, this could not replace the hygienic effect of the ritual washing, especially in times of plagues. Thus, when the Black Death devastated Europe in the years 1348 and 1349, the difference in death rate between the Christian and the neighboring Jewish communities could not be explained in any other way, save by the mischief of the Jews. They were blamed for poisoning the wells from which the Christians drew their water. The frustration of the mobs, facing an unexplained death against which no prayer or sprinkling of holy water could prevail, was released in a wave of executions, mainly by fire, of many Jewish communities.

A small Christian community remained in Jerusalem until the great rebellion of the Jews against the Romans. When Jerusalem was occupied by the fanatic rebels, the Christians either left or were driven out from the city that prepared itself for the siege. The rebellion started in the year 66 A.D. and lasted for 4 years until the conquest and destruction of Jerusalem. A small group of rebels remained in the desert fortresses of Massada and Makhwar in Trans-Jordan which Herod had built. These had been furnished with large stores of water and food, yet these precautions did not help. The two fortresses could not withstand the methodical expertise of the Roman war machine. After 3 years, when the Romans with their siege ramp began storming the walls of Massada , the defenders killed themselves in order not to become enslaved [12].

An investigation of the assemblage of the trees which were used as a basis for the Roman siege ramp [13] showed that the ratio of tamarisk to acacia is higher than that which one would find if one collected such trees at random today. As tamarisk needs a higher groundwater table than the acacia, the botanists who did the survey concluded that the climate was more humid at the time of the siege, a conclusion already expressed in this chapter on the basis of other observations.

In 330 A.D., Constantine the Great built Constantinople as the capital of the Roman Empire on the site of the Greek city of Byzantium. In 395 A.D. the empire was divided between his two sons. During the century which followed, what remained of the western part was devastated by the inflow of West Goths, East Goths, Vandals, Franks, and others. The eastern part survived, was reorganized by the Emperor Justiniane (482-565 A.D.) who also put an end to the Hellenistic-Roman cults, as well as to the school of philosophy of Athens.

In the seventh century A.D. many of the monasteries, strewn all over Anatolia and Greece were deserted. The archeologist Prof. Carpenter [14] maintains that it was not because of raids by Slavic tribes nor Arab pirates, but because a severe phase of aridity. The present author supports this suggestion with the evidence already discussed.

Fig. 12.7. Air photo of Tel Qumran. At the lower left part of the picture the water storage system can be seen (Courtesy Irit Zaharoni, Bamahne Journal, IDF)

13 Epilogue

> A sign there was to Saba in their dwelling places two gardens, the one on the right hand and the one to the left. Eat ye of your Lord's supplies and give thanks to him. Goodly is the country and gracious is the Lord. But they turned aside, so we set them the flood of Irun and we changed their gardens into two gardens of bitter fruit and tamarisk and some few jujube trees. (The Koran, Sura, 34:16-17)

The power of the Roman Empire was founded mainly on its organization. When this disintegrated, the tribes of the north could harass the empire at will. The last successful campaign against them was by Julian the Apostate, who defeated them on the Rhine in 357 A.D. He tried to restore paganism to the Roman Empire and he was tolerant toward the Jews and promised to rebuild the temple in Jerusalem. His plans for putting the old gods of the orient back on the throne did not materialize, as he was killed in his war against the Persians and all his plans for restoring the Roman Empire were buried with him. The tide of the northern tribes rose again. It culminated with the invasion by the Huns and Mongol tribes coming from the vast plains of Asia (370 A.D.). They expelled the West Goths (or Visigoths), who were permitted to cross the Danube and settle in the Roman Empire, where they joined the army and reached high posts in the Roman administration. The movement of the Germans to the south continued. They plundered Athens and Rome, and the Vandals took Spain and established themselves in northern Africa. Was the Hun invasion a function of a climatic change which forced them to look for grazing grounds outside the wide steppes of Central Asia? At this stage of the research, it is difficult to say, as we do not yet have enough paleo-climatic data from Central Asia. The data from the Fertile Crescent and Europe show that, during the fifth century, the climate was becoming generally warmer and thus also dryer. Might this have had an effect also on the steppes of Asia?

As previously discussed, there is no claim whatsoever that climatic changes are the sole cause for the movement of people and historical events; however, that such a connection may have existed is suggested for future research. In other words, it is suggested that future historical research does not adopt a purely anthropogenic explanation as its basic hypothesis.

The same question arises also in regard to the big movement of the Arab tribes which began at the end of the 6th century and reached its climax during the 7th century. It is known that these tribes were driven forward by their new monotheistic religion and belief in Allah. They called themselves Moslems, meaning "the reconciled" and were devoted to spreading the word of Allah by the sword. Yet the timing of the great Arab invasion is very close to the warming phase one can see in the ^{18}O and ^{13}C curve. The evidence from the cores in the ice of Greenland also show warmer temperatures between 400 and 600 A.D. [1]. In the Negev of Israel, the once rich Nabatean-Byzantine cities were gradually abandoned during this period, as well as the monasteries of Greece and Anatolia.

Was this burning religious zeal of the Moslems made fiercer by the droughts which struck the northern and central parts of their peninsula? Did this drying up also weaken the countries of the Fertile Crescent guarding what was left of the Roman Empire, or was it only the natural degeneration and the long and devastating wars with the Persians that caused them to collapse before the Arab invaders?

If northern and central Arabia became dryer, southern Arabia, benefiting from the monsoons, should have become more humid. Archeological investigations have shown that the big dam of Marib was destroyed by a tremendous flood in ca. 575 A.D., 1300 years after it was constructed. It was suggested [2] that this destruction is narrated in the Koran in the words which open this chapter. The Koran, understandably, suggests an anthropogenic reason. Was this the only one?

It is the opinion of the author that it is worthwhile to close this book with a chapter filled with as many questions as possible. The reason for this, as pointed out in the Introduction, is that the purpose for writing this book was not to suggest a new doctrine, but to question the prevailing doctrine of "put all the blame on man".

General Bibliography and References

As this book is a general synthesis, explicit references are made only to the works, the specific findings of which are discussed. The works which helped in the construction of the general background, are mentioned in the General Bibliography.

General Bibliography

Avnimelech, M. A., Bibliography of Levant Geology. Israel Program For Scientific Translations, Jerusalem(1965) I:192pp. (1969) II:184 pp.

Anati, E., Palestine Before the Hebrews. Jonathan Cape, London (1963) 453 pp.

Bar-Yosef, O., Prehistory of the Levant. Ann. Rev. Anthropol. (1980) IX : 1-11

Beydoun, Z.R., The Middle East Regional Geology and Petroleum Resources. Scientific Press (1988) 292 pp.

Bender, F., Geology of Jordan. (Eng. Ed.). In: Contributions to the Regional Geology of the Earth. Martini H.J., (Ed.) (1974) VII : pp.1-196.

Dubertret, L., (ed.) Asie. In: Lexique Stratigraphique Internationale, CNRS, Paris (1960) III: 150 pp.

Dubertret, L., Introduction a la Carte Geologique 1/50000 du Liban. Notes et Memoires . Moyen Orient (1975) 23: 345-403.

Farrand, W. R., Chronology and Paleoenvironments of Levantine Prehistoric Sites as Seen from Sediment Studies. J. Archeol. Sci. (1979) VI:369-392.

Gordon Child, Y., New Light on the Most Ancient East. Percy Lund Humphries & Co., London (1964) 255 pp.

Harel, M., The Ancient Water Supply System of Jerusalem. Geog. in Israel, The Isr. Nat. Com. Int. Geog. Union. Jerusalem. (1976) pp.36-53.

Harden, D., The Phoenicians. A Pelican Book (1971) 313 pp.

Hooke, S. H., Middle Eastern Mythology. A Penguin Book (1966) 199 pp.

Horowitz, A., The Quaternary of Israel. Academic Press, New York (1979) 394 pp.

Josephus Flavius, Jewish Antiquities. Bialik Inst. Jerusalem. (1967) Vols. I & II-624 pp. (Hebrew).

Josephus Flavius, The Jewish War. Harvard University Press. U.S.A.

Marx, A. E. (ed.) Prehistory and Paleoenvironments in the Central Negev. Israel. University of Dallas (1976-1977) I : 383 pp. II : 356 pp.

Mallowan, M.E.L., Early Mesopotamia and Iran. Thames & Hudson, London (1965) 142 pp.

Mazar, B. (ed.) The World History of the Jewish People. Judges. Massada Publ. Co. Israel (1971) III : 366 pp.

Moscati, S., The Face of the Ancient Orient. Doubleday Anchor Books (1962) 375 pp.

Picard, L., Structure and Evolution of Palestine. Geol. Dept. Hebrew Universiy, Jerusalem. (1943)187 pp.

Quenell, A.M., The Structure and Geomorphic Evolution of the Dead Sea Rift. Quart. J. Geol. Soc. London (1958)114 : 1-24.

Said, R., The Geology of Egypt (1962) Elsevier, New York. 377 pp

Wolfart, R., Geologie von Syrien und dem Libanon. In: Beitrage zu Regionale Geologie der Erde. Berlin-Nicolasse (1967) 326 pp.

Woolley, C.L., and Lawrence, T.E., The Wilderness of Zin. New ed. Jonathan Cape, London (1936) 161pp.

Vermes, G., The Dead Sea Scrolls in English. Penguin Books (1970) 253 pp.

Vita-Finzi, C., The Mediterranean Valleys, Geological Changes in Historical Times. Cambridge University Press. (1969) 139pp.

Maps

Bundesanstalt fur Bodenforschung and UNESCO, Carte Geologique Internationale de l'Europe et des Regions Riveraines de la Mediterranee, scale 1:5,000,000 (1971)

Bartov, Y., Eyal, M., Shimron, A. E., and Y.K. Bentor, Sinai Geological Photomap, 1:500,000. Survey of Israel, Tel Aviv (1980).

Bartov, Y.,(ed.) Israel - Geological Map 1:500,000 The Atlas of Israel, Survey of Israel, Tel Aviv (1979)

Commission for the Geological Map of the World, UNESCO, Geological World Atlas, Sheet 9, 1:10,000,000 (1980)

Grollenberg, L.C., The Penguin Shorter Atlas of the Bible. Penguin Books (1978) 265 pp.

Nat. Iranian Oil Co. Geological Map of Iran, 1:250,000 .with Explanatory Notes. (1959)

References Preface

1. Issar, A., Quijano, J.L. Gat, J.R., and Castro, M., The Isotope Hydrology of the Groundwaters of Central Mexico. J. of Hydrology, (1984) 71: 201-224.

References Chapter 1

1. Carpenter, R., Discontinuity in Greek Civilization. Cambridge University Press (1966) 79 pp.
2. Parry, M.L., Climatic Change, Agriculture and Settlement. Dawson Archon Books, U.S.A. (1978) 214 pp.
3. Lamb, H.H., Climate History and the Modern World. Methuen Oxford (1982) 387 pp.
4. Wigley, T.M.L., Ingram, M.J., Farmer, G., (eds). Climate and History. Cambridge University Press (1985) 530 pp.
5. Huntington, E., Palestine and its Transformation. Houghton Mifflin Company, Boston, New York (1911) 443 pp.
6. Huntington, H., Civilization and Climate. Reprint by Archon Books, U.S.A. (1971).
7. Lowdermilk, W.C., Palestine, Land of Promise. Golancz, London (1946) 167 pp.
8. Glueck, N., Rivers in the Desert. Norton & Co., New York (1968) 302 pp.
9. Reifenberg, A., Desert Research. Bull. Res. Counc. Isr. Jerusalem (1953) 3: 378-391.
10. Evenari, M., Shannan, L., and Tadmor N., The Negev : the Challenge of a Desert. Harvard University Press (1971) 345 pp.
11. Jacobsen, T., Salt and Silt in Ancient Mesopotamian Agriculture. Science (1958) 128 : 1251-1259.
12. Issar, A., Geology of the Central Coastal Plain of Israel. Isr. J. Earth Sci. (1968)17 (1): 16-29.
13. Baruch, U., The Late Holocene Vegetational History of Lake Kinneret (Sea of Galilee). Israel, Paleorient (1986)12/2 : 37-48.
14. Stiller, M., Ehrlich, A., Pollinger, U., Baruch, U., and Kaufman, A., The Late Holocene Sediments of Lake Kinneret (Israel)-Multidisciplinary Study of a Five Meter Core. Geol. Surv. of Israel, Current Research (1983-84) pp. 83-88.
15. Issar, A., and Tsoar, H., Who Is to Blame for the Desertification of the Negev? Proc. IAHS Symp. Vancouver, Canada (1987) IAHS publ. no.168. pp.577-583.
16. Issar, A., and Levin D., Climatic Changes in Israel During Historical Times and Their Impact on Hydrological, Pedological and Socioeconomical Systems. Proceedings of the NATO Advanced Research Workshop, (1987).
17. Thompson, F.H., (ed.) Archeology and Coastal Change. The Soc. of Antiq. of London (1980) 154 pp.

18. Davis, O.K., and Turner, R.M., Palynological Evidence for the Historic Expansion of Juniper and Desert Shrubs in Arizona, U.S.A. Rev. of Paleobot. and Palyn. (1986) 49 : 177-193.
19. Geyh, M.A., Wakshal, E., and Franke, H.W., Carbon-14,13 and Oxygen-18 Data of Speleothems from Upper Galilee (manuscript).
20. Bücher, A., Recherches sur les Poussieres Minerales d'Origine Saharienne. These de Doctorat d'Etat. Univ. Reims-Champagne-Ardenne (1986) 165 pp.
21. Wurtele, M.G., The Meteorology of Desertification. In: Progress in Desert Research. Berkofsky, L. and Wurtele, M.G. (eds.), Rowman & Littlefield (1987) 245-260 pp.

References Chapter 2

1. Garfunkel, Z., Pan African (Upper Proterozoic) plate tectonics of the Arabian-Nubian Shield. In: Pre-Cambrian Plate Tectonics. Kroner A. (ed.) Elsevier (1981) 387-405.
2. McClure, H. A., Early Paleozoic glaciation in Arabia. Paleogeog. Paleoclim. Paleoecol. (1978) 25 : 315-326.
3. Freund, R., Garfunkel, Z., Zak, I., Goldberg M., Weissbrod, T., and Derin ,B., The Shear Zone Along the Dead Sea Rift. Phil. Trans. Roy. Soc. London. A: 267 (1970) 107-130.
4. Hsu, K.J., Ryan, W.B., and Cita, M.B., Late Miocene desiccation of the Mediterranean. Nature (1973) 242 : 240-244.
5. Issar, A., see Ref. Chap. 1 [12]
6. Issar, A., Stratigraphy and Paleoclimate of the Pleistocene of Central and Northern Israel. Paleogeog. Paleoclim. Paleoecol. (1979) 29 : 266-280.
7. Korzoun, V., and Drozdov, O.A., (eds.) Atlas of World Water Balance. UNESCO, Paris (1976). 346 pp. 57 maps.
8. Katsnelson, J., Frequency of Duststorms at Beer Sheva. Israel J. of Earth Sciences (1970)19 : 69-77.
9. Issar, A. , and Bruins, J., Special Climatological Conditions in the Deserts of Sinai and Negev During the Late Pleistocene. Paleogeog. Paleoclim. Paleoecol. (1983)43 : 63-72.
10. Tsoar, H., Characterization of Sand Dune Environment by their Grain Size, Mineralogy and Surface Texture. Geography in Israel. Israel National Comittee, Jerusalem.(1966) 327-343.
11. Dan, J., Garson, R., Kouymdjisky, H., and Yaalon D.H., Aridic soils of Israel. Properties, Genesis and Management. Spec. Publ. No. 190. Agricultural Research Organisation, Bet-Dagan, Israel. (1981) 353 pp.

References Chapter 3

1. Kramer, S.N., History Begins at Sumer. Doubleday Anchor Books, N.Y. (1954) 247 pp.
2. Kramer, S.N., Sumerian Mythology. Harper and Row N.Y. (1961) 130 pp.
3. Roux, G., Ancient Iraq. A Pelican Book (1966) 480 pp.
4. King, L. W., (ed.) The Seven Tables of Creation. Luzac and Co., London (1902) Vol.1, 274 pp.
5. Heidel, A., The Babylonian Genesis. University of Chicago Press (1951) 153 pp.
6. Doblhofer, E., Voices in Stone. Collier Books, N.Y. (1971) 327 pp.
7. Woolley, L., Ur of the Chaldees. A Pelican Book (1929-1954) 160 pp.
8. Heidel, A., The Gilgamesh Epic and Old Testament Parallels. University of Chicago Press (1949) 269 pp.
9. Lambert, W.G., and Millard, A. R., Atra-Hasis, the Babylonian Story of the Flood. (Also the Sumerian Flood Story by M. Civil) , The Clarendon Press Oxford (1969) 198 pp.
10. Pritchard, J.B., (ed.) The Ancient Near East, an Anthology of Texts and Pictures. Princeton University Press (1958) 380 pp.
11. Hallo, W.W., and Simpson, W.K., The Ancient Near East, a History. Harcort Brace Jovanovich Inc., U.S.A. (1971) 319 pp.
12. Degens, E.T., Wong, H.K., Kempe, S., and Kurtman, F., A Geological Study of Lake Van, Eastern Turkey . Geol. Rund. Stuttgart (1984) 73, 2 : 701-734.
13. Van Zeist, W., and Bottema, S., Palynological Invetigations in Western Iran. Paleo-Historia (1977)19: 19-85.
14. Hammer, C.U., Clausen, H.B., and Dansgaard, W., Greenland Ice Sheet Evidence of Post-Glacial Volcanism and its Climatic Impact. Nature (1980) 288 : 230-235.
15. Hammer, C.U., Clausen, H.B., Friedrich, W.L., and Tauber, H., The Minoan Eruption of Santorini in Greece Dated to 1645 BC. Nature (1987) 328 : 517-519.
16. Issar, A. , and Levin, D., see Ref. Chapter 1 [16]
17. Reade, J., Assyrian Sculpture. British Museum Publications, London (1983) 72 pp.
18. Cassuto, U., A Commentary on the Book of Genesis. Magness Press, Hebrew University, Jerusalem (1983) (Hebrew) 251pp.

References Chapter 4

1. Picard, L. Baida, U., Geological Report on the Lower Pleistocene Deposits of the Ubeidiya Excavations. Nat. Acad. Sci. Jerusalem (1966) 39 pp.
2. Issar, A., Stratigraphy and Paleoclimates of the Pleistocene of Central and Northern Israel. Paleogeog. Paleoclimat. Paleoecol. (1979-1980) 29 : 261-280.
3. Gilad, D., Hand Axe Industries in Israel and the Near East. World Archeo. (1970) II : 101-133.
4. Ronen, A. (ed.), The Transition from the Lower to the Middle Paleolithic and the Origin of Modern Man. BAR International Series. 151 pp.
5. Issar, A., Fossil Water Under the Sinai Negev Peninsula. Sci. Am. (1985) 253 (I) : 104-112.
6. Issar, A., and Bruins, H., Special Climatological Conditions in the Deserts of Sinai and the Negev during the Upper Pleistocene Paleogeog. Paleoclimat. Paleoecol. (1983) 3.43 : 63-72.
7. Emery, K.O., and Neev, D., Mediterranean Beaches of Israel. Bull. Geol. Surv. Isr. (1960) 26.1 : 1-24.
8. Kenyon, K.M., Digging up Jericho. Benn., London (1957) 272 pp.
9. Mellaart, J., The Neolithic of the Near East. Thames and Hudson, London (1975) 330 pp.
10. Perrot, J., La Prehistoire Palestinienne, un Supplement au Dictionnaire de la Bible (1968) VIII: 416-438.
11. Dothan, M., The Excavations at Naharia. Isr. Ex. J. (1956) 6:14-25.
12. Mazar, B., En-Gedi. In Enc. for Arch. Excav. Isr. Ant. Res. Soc. & Massada, Israel (1970) 440-447 (Hebrew).
13. Bar-Adon, P., The Caves of Nahal Mishmar. In Enc. for Arch. Excav. Isr. Ant. Res. Soc. & Massada, Israel (1970) 354-359.

References Chapter 5

1. Darlington, C.D., The Evolution of Man and Society. Simon and Schuster, N.Y. (1969) 753 pp.
2. Walters, S.D., Water for Larsa, an Old Babylonian Archive Dealing with Irrigation. Yale University Press, New Haven and London (1970) 202 pp.
3. Jacobsen, T., Salt and Silt in Ancient Mesopotamian Agriculture. Science (1958) 128 : 1251-1258.
4. Jacobsen, T., The Waters of Ur, Iraq. Science (1960) 22 :174-185.
5. McNeil, W.H., and Sedlar, J.W. (eds.) The Ancient Near East. Oxford University Press (1968) 261 pp.
6. Neuman, J., Five Letters from and to Hammurapi, King of Babylon (1792-1750 B.C.) on Water Works and Irrigation. J. Hydrol. (1980) 47 : 393-397.

7. Raban, A., Alternated River Courses During the Bronze Age Along the Israeli Coastline. Colloque Internationeaux C.N.R.S. Deplacements des lignes de rivages en Mediterranee. Ed. du C.N.R.S. Paris (1987) 173-189.

References Chapter 6

1. Haldas, A., Who Were the Amorites? E.J. Brill, Leiden (1971).
2. Moscati, S., Ancient Semitic Civilization. Capricorn Books. G.P. Putnam Sons, New York (1960) 254 pp.
3. Albright, W.F., Archeology and the Religion of Israel, Doubleday Anchor Books (1984) 257 pp.
4. Aharoni, Y., The Land of the Bible: a Historical Geography. Burns & Oates (1979) 481 pp.
5. Mazaar, B., Canaan and Israel. Bialik Institute, Jerusalem (1974) 320 pp. (Hebrew).
6. Glueck, N., Rivers in the Desert. Norton & Co. New York (1968) 302 pp.
7. Gvirtzman, G., The Saqia Group (Late Eocene to Early Pleistocene) in the Coastal Plain and Hashephela Regions. Geol. Sur. Isr. Bull. 51,2 (1969) 37 pp.

References Chapter 7

1. Herodotus, The Histories. Translated by Abry de Selincourt. A Penguin Classic (1954) 599 pp.
2. Emery, W.B., Archaic Egypt. A Pelican Book, Penguin Books (1961) 269 pp.
3. Frankfort, H., Ancient Egyptian Religion. Harper and Row, New York (1948) 181 pp.
4. Frankfort, H., Wilson, H.A., and Jacobsen, T., Before Philosophy. A Pelican Book, Penguin Books (1946) 275 pp.
5. Clark, R., Myth and Symbol in Ancient Egypt. Thames and Hudson (1959) 292 pp.
6. Simpson, W.K. (ed.) The Literature of Ancient Egypt, an Anthology. Yale University Press (1972) 347 pp.

References Chapter 8

1. Aharoni, Y., see Ref. Chapter 6.[4]
2. Wilson, I., The Exodus Enigma. Weidenfeld and Nicholson (1985) 208pp.
3. Hammer, et. al., see Ref. Chapter 3[14]
4. Dansgaard, W., Greenland Ice Core Studies. Paleogeog. Paleoclim. Paleoecol. (1985) 50:185-187.
5. Hammer, et.al. , see Ref. Chapter 3 [15]

6. Pachur, H.J., and Roper, H.P., The Libyan (Western) Desert and Northern Sudan During the Late Pleistocene and Holocene. Berliner Geowiss. (1984) A.50 : 249-284.
7. Harel, M., The Exodus Route in the Light of Historical-Geographic Research. In Geography in Israel, 23rd Int. Geog. Cong. USSR (1976) 373-386.

References Chapter 9

1. Ambroggi, R., The Water Under the Sahara. Sci. Am. (1966) 214: 21-29
2. Klitzsch, E., Sonntag, C., Weistroffer, K., and El Shazly, E.M., Grundwasser der Zentralsahara -Fossil Vorrate. Geol. Rund., Stuttgart (1976) 65:264-287
3. Issar, see Ref. Chapter 4 [5]
4. Issar, A., and Gilad, D., Groundwater Flow Systems in the Arid Crystalline Province of Southern Sinai. Hydrol. Sci. (1982) 27: 309-325 .
5. Aharoni, Y., and Amiran R., Arad. Archeology (1964) 17 :43-53.

References Chapter 10

1. Heaton, E. W., Solomon's New Men Pica Press. N.Y. (1974) 203 pp.
2. Kenyon, see Ref. Chapter 4 [8].
3. Issar, A., The Evolution of the Ancient Water Supply System in the Region of Jerusalem. Isr. Expl. J. (1976) 26, 2-3 : 130-136 .
4. Hutchinson, R.W., Prehistoric Crete. A Pelican Book. 367 pp.
5. Finley, M.I., Early Greece, the Bronze and Archaic Ages. Chatto and Windus. London (1981)149 pp.

References Chapter 11

1. Aharoni, Y., The Archeology of the Land of Israel. Westminster, Philadelfia (1982) 344 pp.
2. Simmons, J. J., Jerusalem in the Old Testament by J.E. Brill. Leiden (1952) 517pp.
3. Sasson , J., The Niqba inscription. Pal. Exp. J. (1982) 114 : 111-117.
4. Amiran, R., The Water Supply of Jerusalem. In: Jerusalem Revealed. Shikmona, Jerusalem (1975)134 pp.
5. Issar, A., The Evolution of the Ancient Water Supply System in the Region of Jerusalem. Isr. Expl. Jou. (1976) 26,2-3 :130-131.
6. Sandstrom, G.E., Man the Builder. McGraw-Hill Books (1970) 280 pp.
7. Pritchard, J.B., Ancient Near Eastern Texts Relating to the Old Testament. Princeton (1955) 380 pp.
8. Arberry, A.J. (ed.), The Legacy of Persia. Oxford Press (1953) 419 pp.

9. Finegan, J., The Archaeology of World Religions. Princeton University Press (1952) I : 599pp.
10. Doblhofer, E., Voices in Stone. Collier Books, New York (1961) 326 pp.

References Chapter 12

1. Josephus Flavius, Jewish Antiquities. See general Bibliography.
2. Adams, F.D., The Birth and Development of the Geological Sciences. Dover Publications (1938) 566 pp.
3. Glueck, N., see Ref. Chapter 6 [6].
4. Evenari, M., et. al., see Ref. Chapter 1 [10].
5. Hammer, et. al. see Ref. Chapter 3 [14,15]
6. Klein C., Morphological Evidence of the Lake Level Changes, Western Shore of the Dead Sea. Isr. J. Earth Sciences (1982) 31:67-94
7. Nicholson, S.H., and Flohn, H., African Environmental and Climatic Changes and the General Circulation in the Late Pleistocene and Holocene. Climatic Change (1980) 2(4):313-348.
8. Thompson, F.H. (ed.) Archeology and Coastal Change. The Society of Antiquitaries of London (1980) 154 pp.
9. Shanan. L., Rainfall and Runoff Relationships in Small Watersheds in the Avdat Region of the Negev Desert Highlands. Ph.D. Thesis. The Hebrew University, Jerusalem (1975).
10. Yaalon, D.H., and Dan, J., Accumulation and Distribution of Loess-Derived Deposits in the Semi-Desert and Desert Fringe Areas of Israel. Z. Geomorph. N.F.Suppl. (1974) 20 :41-61.
11. Issar, A., Nativ, R., Karnieli, A., and Gat, J., Isotopic Evidence of the Origin of Groundwater in Arid Zones. Isotope Hydrology, IAEA/UN, Vienna (1984) 85-104 pp.
12. Josephus Flavius, The Wars of the Jews Against the Romans. See General Bibliography.
13. Liphschitz, N., Lev-Yadun, and S., Waisel, Y., Dendochronological Investigations in Israel (Masada). Isr. Exp. J. (1981) 31: 230-234 .
14. Carpenter, see Ref. Chapter 1[1]

References Chapter 13

1. Hammer et . al. , see Ref. Chapter 3 [14,15]
2. Evenari, M., Desert agriculture past and future. In: Settling the Desert, Berkofsky, L., Faiman, D., and Gale, J. (eds.) Gordon and Breach (1981) 3-28.

Appendices

I Principles of Well Construction

In order to understand the technique of digging, and constructing wells in arid zones, a short discussion on the principles of groundwater flow has first to be given.

The source of all freshwater on earth is the sea, where vapor is extracted by the impact of energy from the sun and transported toward the land by the energy of winds. There the masses of air containing water vapor may go through a cooling process, for instance when they meet a barrier of a mountain or a cold front and are driven upward. They can also be caught in an ascending whirlpool due to the overheating of the lower layer of air by the hot surface of earth. The ascending air becomes cooler and thus also becomes more saturated. Due to this process droplets of liquid or even ice may form. At a certain moment the force of gravity acting on these droplets or ice crystals becomes strong enough to cause them to fall as rain, hail, or snow.

After reaching the ground, this water becomes subject to numerous forces or energy fields. Firstly, thermal energy works on it, causing the water to evaporate. In arid zones the rains may even evaporate before reaching the ground. It is a common sight in deserts on a warm cloudy day to see sheets or veils of rain come down from a cumulus cloud but not reach the ground.

Another field of force is that of gravitation, which pulls the water to the lowest point on earth. If the ground is porous the water will infiltrate into the subsurface. When the rock is permeable but the rain is not intensive enough, all the water will be absorbed by the soil. Thus, in areas of sand dunes, one can walk in a rainstorm without the soles of one's shoes becoming soaked. If the ground is impermeable or partly permeable and cannot transmit all the rain water to the subsurface, the water will accumulate on the surface to form runoff. The water will flow over the surface to form streamlets which flow to the creeks and river-beds to form floods. In deserts built of impermeable rock where no pervious absorbing soil exists, the floods may form almost immediately after the rain starts, flowing in large volumes, then stopping abruptly shortly after the rainstorm terminates.

The water absorbed in the subsurface immediately after the rains, or from the water of the floods running over permeable gravel layers, starts its voyage in the subsurface in a vertical direction through the pores, fractures, and channels in the soil or rock until it reaches an impervious layer at a shallow or deep level. In the upper layers, where roots are active, the water may be drawn back upwards by the roots, and be transpirated. Some trees like the acacia may have roots a few meters deep. The water which does not transpirate accumulates in the subsurface on the impervious layers. The pores and fractures above these layers become saturated. The water in the saturated rock is termed groundwater. The direction of movement in this body is semi-horizontal. The plain which separates this body from the overlying nonsaturated rock is called the water table. The water table slope or gradient indicates the direction in which the groundwater will flow. This gradient is very low, a few meters per kilometer and the flow is very slow, from a few to a few hundred meters a year. The reason for this can be grasped if one can imagine the subsurface as composed of billions of minute retarding dams which cause the water molecules to slow down. Only under special circumstances, mostly in limestone rocks, do the solution channels become big enough to allow a rapid flow of groundwater and only in such cases may subterranean rivers flowing in large subsurface caves be formed.

In places where the water table meets the surface, either in a channel of a river or along a cliff formed by a fault, a spring emerges. The quantity and the duration of flow is a function of the permeability of the rock and quantity of the water stored in the saturated layers.

As already recounted, the earliest human communities in the Levant established themselves near such springs. For example, the water of the spring of Jericho furnished the supply to one of the earliest irrigation projects. In many aspects the utilization of spring water is easier than that of river water, once the reliability of its flow has been established.

Springs in semi-arid and arid zones, however, are rather scarce, and their water may not suffice to supply the demands of an expanding population. Thus, when the people had to leave the overcrowded spring and river valleys, they had to locate another source of water for themselves and their livestock. The needs of men are rather small and often the water accumulated in a pothole in a river-bed or in a hole excavated into the impervious ground sufficed for a hunting party or a traveller crossing the desert from one watered valley to the other. But after goats, sheep and cows, and the beasts of burden and transportation like the donkey, horse, and camel were domesticated and the forage in the vicinity of the spring or river was not sufficient, men had to look for a dependable and adequate supply of water away from the spring. It is natural that he went looking for the water hidden underground.

It is quite obvious that the first wells were dug in river-beds in which the flow had dried up. It is quite natural even for a primitive man, seeing the water disappear, and seeing the moist soil, to excavate a hole to find the disappearing resource. In the deserts of Namibia elephants were observed to dig into the soil which was drying up to locate water. Thus it can be assumed that the techniques of well excavation were first developed in the ephemeral river-beds.

The excavation to a small depth is not a difficult task and even primitive societies can maintain themselves along river-beds which continue to carry water in the subsurface during the dry seasons. Today this technique is used by bedouins in the deserts of Sinai, and this type of shallow well is called a Tamila.

When the water table below the river-beds drops, reaching the supply of water becomes more difficult, as the excavation has to go deeper. The walls have to be protected and the material which is excavated has to be carried up from deep below the surface.

In the deserts of the Levant one can find all stages of the evolution of the techniques of excavation. The nomads satisfy themselves by a Tamila or a deep hole which they excavate, hauling out the material by a rope and basket. The walls of the hole are seldom protected. The digger goes down into and out of the hole by digging small holes in the walls of the shaft. The author has witnessed such well-digging in the wadis of southern Sinai, where the diggers used this method to excavate down to 20 meters. If they did not feel safe or a deeper hole was needed, the walls were supported by boulders built around the wall. In such cases a tripod made of tamarisk or acacia trunks is built above the shaft hole and a camel drawing a rope and basket helps in the operation.

When the well reaches the water table, the diggers dig into the saturated zone, removing the water with a bucket which is drawn to the surface by the rope. When the quantity of water to be removed by the rope and bucket does not suffice to empty the hole and water reaches the loins of the diggers, the digging comes to a stop and the well is finished. Thus the depth of the bottom of the well below the water table can tell the surveyor about the capacity of an ancient well. The deeper such a well is below the water table, the smaller is its capacity.

In order to protect against collapse of the walls, especially below the water table and when layers are friable, various techniques of wall building were adopted. During Roman-Byzantine times, the building of masonry protecting walls from the surface to the bottom was developed. In this way, the well-digger progressed downward by underlying each row of masonry with another, leaving openings between the stones for the water to seep in.

General Bibliography

1. Todd,K.T., Ground Water Hydrology . Wiley N.Y. (1959).
2. Johnson E.E. Inc. Ground Water and Wells. Johnson Inc. St. Paul Minnesota, (1966).
3. Intermidiate Technology Development Group Ltd. Hand Dug Wells. London (1976).
4. F.A.O./U.N. Self-help Wells .Food Agriculture Organisation, Rome. Paper no. 30 (1977).

II The Environmental Isotopes, Deuterium and ^{18}O, the "Finger-Prints" of Water

The method of isotope analysis had been applied to investigate the fossil water resources of the Negev and Sinai. In this method, amounts of heavy isotopes of hydrogen and oxygen in water (deuterium = ^{2}H and ^{18}O) are measured. Unlike the ^{14}C radio-isotope, the ^{18}O and deuterium isotopes are stable; they do not undergo radioactive disintegration. These isotopes, unlike ^{14}C and tritium, are not produced by cosmic radiation. They have been part of the earth's hydrosphere since the Big Bang. Due to differences in the intermolecular bonds, molecules containing ^{2}H and/or ^{18}O will be involved in the process of evaporation to a lesser degree than the lighter water composed only of $^{1}H_2\,^{16}O$. Therefore, the relative quantity of heavy isotope in water evaporated from the sea, forming clouds and rain, is less than that of ordinary ocean water. Taking the standard mean ocean water (SMOW) as the starting point, and plotting the composition on a chart (Fig. II.1), this atmospheric water will have a negative sign in relation to the composition of ocean water (SMOW).

Due to thermodynamic reasons, there will be eight times more ^{2}H water molecules in the air than those composed from ^{18}O. This relation will change according to the physical environment in which the evaporation takes place. The relative difference between the ocean and the atmosphere regarding heat saturation and transfer rate will also affect the relative abundance of ^{2}H and ^{18}O. These relations are expressed in the following equation:

$$^{2}H^{o}/oo = 8\ ^{18}O^{o}/oo + d,$$

where $^{o}/oo$ expresses the relative difference from the Standard Mean of Ocean Water (SMOW) per mille, and +d, also called the deuterium excess, expresses the local variation due to the local conditions of evaporation.

A network of rain-sampling stations distributed all over the world by the International Atomic Energy Agency have established the fact that on the average, the world's rainfall has a deuterium excess of about 10. In the Mediterranean area, on the other hand, the deuterium excess of contemporary rains is about 22. Thus, if the ratios of $^{18}O^{o}/oo$ in rainwater are plotted on a chart with the ratio in SMOW as the zero point (Fig. II. 1), the rains in countries surrounding the oceans will fall on a line described by the equation $D^{o}/oo = 8\ ^{18}O^{o}/oo$. This is called the "global meteoric line". The rains in the countries affected by the Mediterranean will fall on a line with the equation $D^{o}/oo = {}^{18}O^{o}/oo+(\sim)22$, called the "Levant meteoric line".

The arrangement of the rain samples along the meteoric lines will also be a function of the site of the rain event in relation to its distance from the oceans and its altitude. The rains nearer to the oceans will be heavier and closer to SMOW, while the further from the oceans the rain event or the higher the altitude the storm is, the lower its composition in ^{18}O and deuterium which will be plotted on the meteoric lines. These shifts to lighter water are called continental and altitude effects.

The rainwater reaching the surface of the land will again undergo the process of evaporation. This will further affect the water's isotopic composition, causing it to become heavier in these isotopes. However, the water which infiltrates immediately to the subsurface to form groundwater will escape further evaporation. In most cases the infiltrating water was found to represent the average composition of the precipitation over the groundwater replenishment area, except in areas of geothermal anomalies at depths where temperatures are above 80°C. In these areas, an isotopic exchange of ^{18}O will occur, and the water will become enriched in ^{18}O.

Thus it can be seen that the isotopic composition of water serves as a very good clue to the environmental conditions under which the water was recharged, especially when this information is used in combination with the ordinary chemical analysis of the water, i.e., the quantity and ratio of the salts dissolved in the water. The water's age is determined either by the tritium content (if the water is less than 40 years old) or by ^{14}C if not older than 40,000 years.

The impact of the use of isotope analysis on the understanding of the hydrogeology of the Sinai and the Negev is now obvious. The analysis of water samples brought from Uyun Musa, those from an abandoned oil exploration well dug into Nubian sandstone in a place called Nakhl in central Sinai, and samples from artesian wells in the Nubian sandstone layer near the Dead Sea, all showed the same chemical and isotopic character. Also, their ^{18}O and deuterium ratios fell around the global meteoric line, while contemporary rains and the groundwater of Sinai and Israel all fell on the Levant line (Fig. II.1). These results support the hypothesis that the Sinai and Negev are underlain by the same regional aquifer, having its outlets along the regional fault lines of the Suez and the Dead Sea Rift Valleys, bordering these areas to the east and the west, respectively. The results also support the hypothesis that the climate during the infiltration of this water was different from the contemporary one.

^{14}C dating gave the age of about 20,000 years for water from the Nakhl well and 30,000 years for water from Uyun Musa and the Dead Sea region. These match the results obtained from the calculations of groundwater flow velocity from the oil wells mentioned above to the outlet areas. The calculations were based on the hydraulic gradients, taking into account the hydraulic permeability coefficient. These ages are similar to those determined for the water in the Sahara and coincide with the latter part of the Last Glacial age which dominated our globe from approximately 80,000 years B.P. to 15,000 B.P.

General Bibliography

1. Fritz, P., and Fontes, J. Ch., Handbook of Environmental Isotope Geochemistry. Vol. 1. Elsevier, Amsterdam (1980).
2. Ferronsky. V. I., and Polyakov. V. A., Environmental Isotopes in Hydrosphere. John Wiley and Sons (1982) 466 pp.
3. Gat, J., Stable Isotope Hydrology, Deuterium and Oxygen-18 in the Water Cycle : A Monograph. IAEA, VIENNA (1981) 339pp.
4. Hoefs, J., Stable Isotope Geochemistry, Springer-Verlag, (1980) 208pp.

5. IAEA (International Atomic Energy Agency) Isotopes in Hydrology, (proceedings Series) , Vienna. (1970,1974, 1978,1983)
6. Arid Zone Hydrology : Investigations with Isotope Technics, Vienna (1980) 125pp.

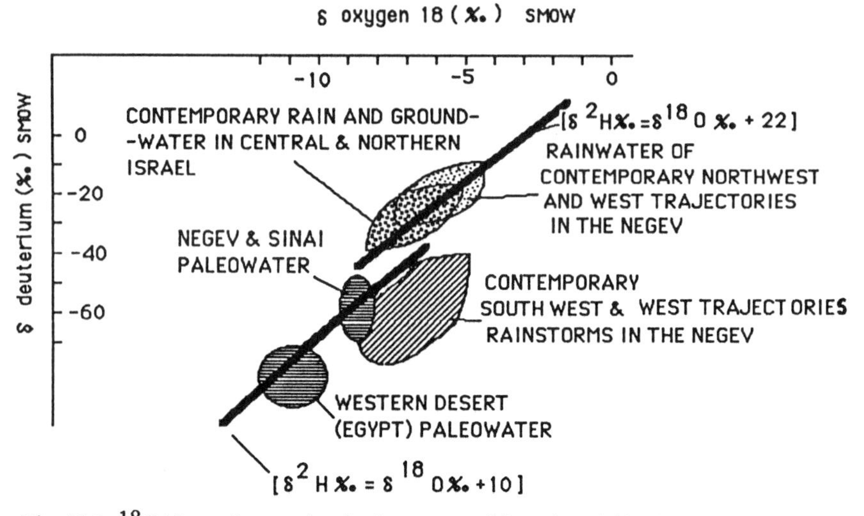

Fig. II.1. ^{18}O/deuterium ratios in the water of Israel and Sinai

III The Sulfate Anomaly

Before the heavy isotopic composition supplied convincing evidence of the similarity between the water found along the Arava and the Dead Sea Rift Valley borders and that of Uyun Musa in the Suez Rift Valley, the chemical composition of the water in the Nubian sandstone aquifer was used as a tracer especially, the anomaly of a sulfate content was found in all cases.

This anomaly was a source of unpleasantness on many of the author's field trips to the Arava and Sinai during the early stages of exploration. Since the water found in the area was not yet desalinated, on hot summer days when much water had to be consumed, it was like taking a large dose of Epsom salts; this, in addition to the need to overcome the bitterish-salty taste of the water.

What is the reason for this sulfate anomaly? Many working hypotheses were formulated; however, they were unacceptable since they did not comply with the basic observation that sulfates, either gypsum or epsomite, are seldom found in the sandstone outcrops of Lower Cretaceous age in which this water is found. During the last years some interesting observations have been collected which point toward an answer. The solution is found with the aid of the environmental isotopes, namely, the deuterium and ^{18}O ratios.

Hydrogeological research of the Negev carried out by the author's team has put special emphasis on the study of the chemical and isotopic composition of rain and floodwater. This was in order to try to promote new tools for the understanding of the hydrogeological and chemical processes which water undergoes in the desert. The difficult conditions of the terrain, the scarcity of settlements, the randomness of rainstorms and flood events, together with the requirement of taking nonevaporated samples, led the research team to develop a special sampling device. This sampler is composed of a funnel, the diameter of which was calculated in such a way as to yield 200 centiliters of water, the quantity needed for a complete chemical and isotopic analysis. In each sampler is a magnetic device which hermetically seals the container after it is full. The waters are then diverted to the next container. These samplers were places in observation stations all over the desert. In order to sample floodwater which infiltrates into the gravel beds of desert streams, samplers were placed in special trenches dug into the stream beds. The samplers were then covered with gravel [1].

After a rainstorm and subsequent flood events, the samples of water were collected and analyzed for their chemical and isotopic composition. The results of this analysis revealed interesting results regarding the chemistry of the rains and floods. One finding was that the quantity of airborne salts brought in by the rainstorms was larger than the quantity exported by the floods. This explains, experimentally, the process of salinization of desert soils. A comparison between the trajectory of a rainstorm and the rains' isotopic composition showed a statistical correlation between the isotopic composition and the direction from which the storm came. Most of the rainstorms come from the northwest and north and are similar in isotopic composition to the rains of the northern part of Israel. They have a deuterium excess of about 22°/oo. The few storms that come from the west and southwest are depleted in deuterium and ^{18}O and are characterized by a deuterium excess of 10°/oo[2].

The reason for the similar isotopic composition of the water of these rainstorms coming from the south and the southwest and that of the paleowater of the Nubian sandstone can be explained by the thermodynamic conditions prevailing over the Mediterranean, from which these storms receive most of their moisture, and the route which they follow over the deserts of Libya and Egypt. This conformity in isotopic composition is a clue to the nature of the climatic conditions which prevailed in this part of the world during the Last Glacial period when the Nubian sandstone aquifers were filled.

It is suggested that during this period, a cyclonic cold high pressure area over Europe caused a southerly deflection in the route of many rainstorms coming from the northern Atlantic and travelling over the Mediterranean. They took a more southerly path and entered the African continent over the Libyan coast and crossed the Sinai and the Negev from the west and southwest. Travelling over a relatively warm Mediterranean during the autumn and possibly during the summer, the water in these storms became similar in isotopic composition [3].

Now for the explanation of the sulfate anomaly. The bathimetric chart of the Suez Gulf and the coast of the Mediterranean show that the depth at these places does not exceed a few tens of meters. Since during the Last Glacial period the sea retreated to a depth of about 100 meters, one can conclude that during such a time, the Gulf of Suez and the upper shelf area of the Mediterranean coast along the Sinai turned into Salinas, or sabkhas, similar to the Sabkha of Bardawil which stretches along part of the Sinai coast today. Evidence for such a retreat of the sea was found in the profiles of wells drilled along the coastline. Storms blowing from the sea over these sabkhas will carry a load of salts, especially gypsum. When they let down the moisture they carry as precipitation, the salts will infiltrate the sands to enrich the groundwater with some table salt and especially with gypsum, namely, $CaSO_4$ [4, 5].

References

1. Adar E., Levin, M., Barzilai, A. Development of Self-Sealing Rain Sampler for Arid Zones Water Resources Res.(1980)16 : 592-596.
2. Leguy, C., Rindsberger, A., Zangvil, A., Issar, A. and Gat J., The Relation Between the Oxygen-18 and Deuterium Content of Rainwater in the Negev (Israel) and Air Mass Trajectories. J. Of Isotope Geosciences. (1983)I 205-218.
3. Issar, A. and Gilead, I., Pleistocene Climates and Hydrology of The Negev (Israel) and Sinai (Egypt) Deserts. Berliner Geowiss. Ab. (1986) 17-25.
4. Issar, A., Fossil Water under The Sinai Negev Peninsula. Sci. Am., (1985) 253:104-111.
5. Issar, A., Bahat, D., and Wakshal, E., Occurrence of Secondary Gypsum Veins in Joints in Chalks in the Negev, Israel. Catena (1988) 15 : 241-247.

Index